Principles and Methods of

PLANT MOLECULAR BIOLOGY, BIOCHEMISTRY AND GENETICS

Dr. Pratibha Devi

M.Sc., Ph. D.

Plant Biotechnology Lab, Department of Botany

Osmania University, Hyderabad, 500 007 (India)

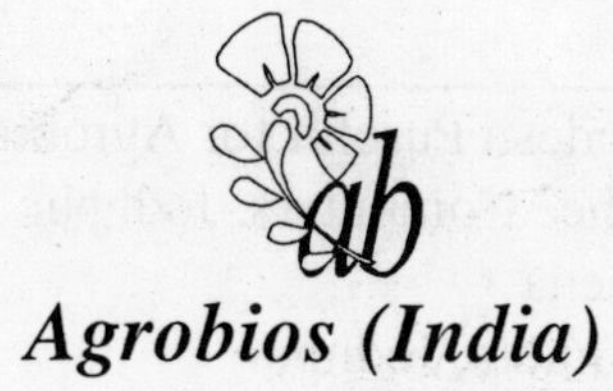

Agrobios (India)

Published by:
Agrobios (India)
Agro House, Behind Nasrani Cinema
Chopasani Road, Jodhpur 342 002
Phone: 91-0291-2642319, Fax: 2643993
E. mail: agrobios@sify.com

Agrobios (India)

First Published 2000
Reprinted : 2002
Reprinted: 2005

ISBN No.: 81-7754-051-3

Price: Rs. 495.00/ US$ 49.00

Published by: Dr. Updesh Purohit for Agrobios (India), Jodhpur
Lasertypeset at: Yashee Computers, Jodhpur
Cover Design by: Reena
Printed at: Bharat Printers, Jodhpur

Dedicated to

My Beloved Parents

Foreword

During the one year that Dr. Prathibha Devi lived in Michigan and worked at Michigan State University, USA, I had the opportunity to read several drafts of sections of this book. In my eyes, the compilation of techniques and experimental protocols for Plant biochemistry and Molecular Genetics appeared to be unique in its breadth. And as such, it was a very ambitious undertaking. As I read the chapters that Dr. Devi brought me, I came to see what a wonderful resource the experimental protocols would be for students who must step beyond the boundaries of their mentors' experiences.

This book would benefit students who take up Biochemical studies, Cytological and Genetical analysis and who prepare gene constructs, manage to develop transgenic plants and find the need to characterize them for the presence of genes and so on. Although many of the methods can be followed as one would a cookbook, I am sure that the inexperienced researcher will appreciate the explanations that punctuate the protocols. I am happy to represent the Plant Science community in saying that Dr. Devi's efforts are appreciated. Her book will have a welcome place on my own bookshelf, although I intend to loan it out frequently.

Barbara B. Sears
Professor
Department of Botany and Plant Pathology
Michigan State University
East Lansing, MI 48824, U.S.A.

Preface

This book was written to help researchers, teachers, and students to properly understand and use the major techniques and methods employed in cellular and molecular biology. Protocols for some of the experiments are no doubt present dispersed in various books and scientific journals, but these are not always accessible to a scientific worker. My objective was to bring all related protocols together within a framework of six chapters (and of course the appendix) so that they are readily available and accessible to the average scientific worker. This material should be useful to the under-graduate as well as post-graduate students in colleges and universities who deal with Cell Biology and Biochemistry as well as Genetics and Molecular Genetics as parts of their courses. This book extensively deals with the Plant Sciences and the experiments can be easily tailored to suit individual conditions, hence should be of general interest to researchers and teachers who frame the syllabi.

Each chapter covers an outline of theoretical concepts and principles of various protocols followed by a detailed practical methodology. Aspects on General Laboratory Principles together with a brief description of various instruments and methods are dealt with in the first chapter followed by the second chapter on qualitative and quantitative estimations in Plant Biochemistry and physiology. The chapter on Cytology concerns crisp methods for cytological study, followed by the chapter on Genetics, which includes theory and problems besides statistical analysis. The chapter concerning Plant Molecular Genetics includes a sizeable collection of protocols dealing with various aspects of recombinant DNA technology and gene cloning. The sixth chapter deals with Plant Genetic Engineering, which includes tissue culture methods followed by protocols of gene transfer and molecular analysis of the putative transformants.

The experiments in this book are the result of a number of years of experience in teaching under-graduates, post-graduates and guiding research scholars for the Ph.D. degree. Their queries and interest have been the main inspiration for writing this book. Further, the practical experience gained by the author during the one-year stint as a visiting scholar at the Genetic Engineering Laboratory of Dr. Mariam Sticklen, Michigan State University, East. Lansing, MI, USA, from September 1997 - 98 has helped in thoroughly revising the book with respect to the protocols. I am grateful to her for providing me this opportunity. Dr. Barbara Sears, Professor, Department of, Botany and Plant Pathology, Michigan State University, MI, USA has critically reviewed the manuscript painstakingly for more than six months and offered several suggestions to bring it to International standard. I acknowledge her with immense gratitude.

Many teachers and colleagues in the Dept of Botany, Osmania University, Hyderabad, who have helped me over the years in gaining thorough knowledge of the various aspects of Biology dealt in this book have contributed immensely towards this effort. I am indebted to them. I am also particularly indebted to the research scholars who have helped me design and test most of the experiments. I am thankful to the faculty members of Centre for Plant Molecular Biology, Osmania University, for their help, especially Dr. K. K. Jena (presently with MAHYCO) for offering valuable suggestions.

I am thankful to late M. Satish Kumar and B. Srinivas, M/S MS Computers, Hyderabad, for helping in the computerization process. Finally, I thank my family members for their encouragement, understanding, tolerance and help without which I could not have accomplished this task.

Suggestions for improvement of this book are most welcome.

Prathibha Devi

Hyderabad

Contents

CHAPTER: 2
PLANT BIOCHEMISTRY AND PHYSIOLOGY 31

CHAPTER: 5
PLANT MOLECULAR GENETICS 139

CHAPTER: 6
PLANT GENETIC ENGINEERING 201

APPENDIXES

List of Figures

1

General Laboratory Principles

The researcher should read and understand the protocol of an experiment before actually starting it. Interpretation of results and trouble-shooting require careful maintenance of experimental records. An important necessity is to always include controls (a negative and a positive control) in an experiment. Each researcher should be aware of the safety considerations of the laboratory. Essential laboratory skills include the ability to prepare solutions and buffers and practical knowledge of the principles of the instruments and techniques used routinely in a biology laboratory. This chapter covers most of such basic general principles of biology.

LABORATORY RECORDS

Any laboratory experiment should include maintenance of clear and accurate records of all experiments conducted. It is important to record all experimental findings in a notebook with dates and page numbers. The first few pages can be used for a table of contents. All the details for each experiment can be organized as shown below:

1. Brief title of experiment and date.
2. Aim.
3. Procedure: Methodology or a flow chart. Each experiment should ideally have three replicates and appropriate controls (positive and negative controls).
4. Results: Presented in Tabular form or as Graphs.
5. Analysis of data: The average values are calculated from the replicates and noted.
6. Inference: The results obtained should be interpreted in accordance with the principle of the experiment.

LABORATORY REQUIREMENTS AND SAFETY

The basic requirements of a biology laboratory are: A good computer and printer, autoclaves, sinks, water distillation units, deionizers, hot-air ovens, chemical fume hoods, radiation shields, sterile laminar flow workstations, incubators, temperature control or cold room, dark room for development of photographic films, -20 °C freezers, refrigerators, water baths, refrigerated centrifuges, micro centrifuges, pH meter, scintillation counters to monitor radioactivity, trans-illuminator equipped with a camera (to view and photograph gels under UV light) or electronic imaging system, electrophoresis units, blotting apparatus, hot plates, micro wave ovens, balances and many other miscellaneous things.

The laboratory needs a variety of glassware including reagent bottles, beakers, measuring cylinders, Ehrlenmeyer (conical) flasks, test tubes, burettes, pipettes, glass rods, Petri dishes and other culture vessels. Disposable sterile Petri dishes can be used for the culture of microorganisms and plant tissue. Magenta boxes and other autoclavable containers can be used for the culture of plants in *vitro*. Other requirements include liquid nitrogen containers, syringes, needles, forceps, scalpels, membrane filters (to filter sterilize heat labile liquids), magnetic stirrers, orbital shakers, inoculation loops, stop watches, nitro cellulose or nylon membranes, parafilm, saran wrap, aluminium foil, marker pens, Whatman 3 M paper, ice buckets, latex gloves, plastic boxes, plastic bags and U.V goggles, besides plastic bottles and containers. A good supply of chemicals is a very essential aspect of equipping a laboratory as are uninterrupted power and water supply.

Adequate safety measures should be taken to protect the workers like providing safety hoods, good radioactive waste disposal systems, gloves when using hazardous carcinogenic chemicals and wearing of goggles for protection from UV light. Precaution should be taken for containment of pathogenic organisms by autoclaving the culture tubes and glassware before washing them. Other contaminated materials should be rinsed in Chlorox solution for a few minutes before washing them. Working behind a radioactive protection shield when dealing with high-energy radioisotopes such as ^{32}P is mandatory. The work area should be checked for spills by using a Geiger Muller counter.

Exposure to several hazardous and toxic chemicals and other agents in a laboratory poses some danger to the researcher, and it is essential to adopt safety measures for protection:

1. Never use equipment or a chemical without reading the information and instructions.
2. Read the warning signs or labels on equipment & chemicals before using them. Make sure you know the location of the safety equipment like the eyewashes, first aid kits, cleanup kits and fire extinguishers and learn how to use them.
3. Wear lab coats, gloves, eye protection and inhalation protection masks when working with chemicals, UV light etc.
4. Work with volatile or potentially hazardous chemicals in a laboratory fume hood only.
5. Take extreme precaution to dispose of the radioactive wastes in appropriate containers and clean up spills immediately. Check before and after working in an area of the laboratory, with a Geiger Müller counter for any radioactive contamination. If any radioactivity resists, the problem should be reported to the concerned authorities and efforts made to remove it.
6. In case of an injury, seek medical aid immediately.

LABORATORY TECHNIQUES

The most important laboratory techniques include the preparation of solutions, measurement of pH, preparation of buffers etc.

Preparation of solutions

Solutions can be prepared by using different units of measurement.

Concentration in percentage

i. Percent by weight (w/w; g "solute" /100 g)

ii. Percent by volume (v/v; ml/100 ml)

Percent by weight per volume (w/v; g "solute" /100 ml)

Concentration in molarity

$$\text{Molarity} = \frac{\text{Number of moles of solutes}}{\text{1 litre solution}} = 1\text{M}$$

i.e the molecular weight of the solute is dissolved in a solvent to make one litre of the solution 1M.

Preparation of buffers

The pH value describes the acidity or alkalinity of solutions. It is the hydrogen ion concentration of a solution. A buffer is a solution that resists pH change upon addition of acid or base to it. Buffers are used to neutralize the acidic or basic products of biological reactions *in vitro* and maintain the pH near the physiological range. Buffers consist of a mixture of a weak acid or base and its salt e.g. acetic acid and sodium acetate. Some of the more common buffer solutions used in the laboratory are given below.

Phosphate buffer (0.1M)

Stock solutions

A. 0.2 M solution of $Na_2HPO_4.7H_2O$ = 35.61 g in 1 L dH_2O.

B. 0.2 M solution of $NaH_2PO_4.2H_2O$ = 31.21 g in 1 L dH_2O.

To obtain the required pH, mix stock solutions (as shown in table 1) and dilute to 200 ml with dH_2O.

Table - 1

pH	[A] 0.2 M Na_2HPO_4 (ml)	[B] 0.2 M NaH_2PO_4 (ml)
6.0	12.3	87.7
6.4	26.4	73.5
7.0	61.0	39.0
7.4	81.0	19.0
8.0	94.7	5.3

1 M Tris - HCl Buffer: [(hydroxymethyl) amino methane]

Dissolve 121.1 g of Tris base in 800 ml of dH_2O. Adjust the required pH with concentrated HCl and bring up to 100 ml. The details are given below:

Table - 2

Tris (g)	dH_2O (ml)	HCl added to Tris (ml)			Final volume (ml)
		PH 7.4	pH 7.6	pH 8.0	
121.1	800	70	60	42	1000
60.55	300	35	30	21	500
12.11	75	7	6	4.2	100

0.5 M Tri

Dissolve 121.1 g Tris in 1.5 L dH_2O, adjust the pH with 4N HCl and make up to 2L with dH_2O.

0.05 M Tris

Dilute 100 ml of 0.5 M Tris to 1L with dH_2O and check the pH.

T E buffer: (10 mM Tris 1mM EDTA with pH 8.0)

Mix 1 ml of 1M Tris, pH 8 and 1 ml of 0.1 M EDTA and make up to 100 ml with dH_2O. (To prepare 0.1 M EDTA, dissolve 37.2 g EDTA (disodium salt) in dH_2O and make up to 1 L with dH_2O.

10 X TBE buffer: (Tris-borate EDTA buffer, pH 8.0)

Dissolve 216g of Tris and 110 g boric acid in 1.2 L H_2O. Add 400 ml 0.1 M EDTA and adjust pH with Tris buffer or boric acid. Dilute to 2L with dH_2O. To prepare 1X TBE buffer, dilute 100 ml of 10X TBE to 1 L.(i.e. dilute 10 times).

50 X TAE Buffer: (Tris-acetate-EDTA buffer, pH 8.0)

Dissolve 242 g Tris in 57.1 ml glacial acetic acid and 100 ml of 0.5M EDTA. Make up to 1L with dH_2O. Dilute 50 times to prepare 1X TAE buffer.

USE OF EQUIPMENT

Use and calibration of micropipettes

Micropipettes are one of the most important tools of a molecular biologist. Micropipettes of different brands have their own unique features. However, the general features and the method of working are all the same (Fig-1). Micropipettes need to be used with tips. The tips (which can be sterilized by autoclaving) come in two different sizes; the smaller tips are used with the P-20 to P-200 micropipettes and the bigger ones with the P-1000 type. The desired volume can be adjusted by dialing the ring on the micropipette. The capacity and range of P-20 is 0 - 20 ml; P- 100 is 0 - 100 ml; P - 200 is 0 - 200 ml and: P - 1000 is 0 - 1000 ml.

Caution !

- Set the volume only within the range of the micropipettes.
- Fix the tip firmly by pressing the pipette into it. Change tips for each new reagent and sample.
- Always hold a micropipette in a vertical position when the liquid is in the tip to prevent it from running back into the nose cone.

- ➢ Practise depressing the plunger with the thumb so that you notice the first and the second stops.
- ➢ Regulate the plunger and take up or eject the fluid with gentle movements.

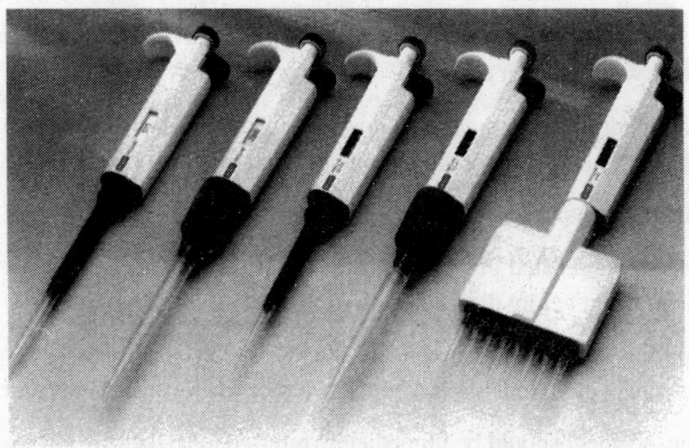

Fig: 1: Micropipettes

Taking up the sample

1. Hold the micropipette vertical and keep it and the container with the fluid almost at eye level.
2. Depress the plunger of the micropipette to the first stop and dip the tip into the liquid.
3. Draw the fluid into the tip by slowly releasing the plunger.

Expelling the sample

1. Open the lid of the tube into which the sample needs to be ejected.
2. Position the tip of the micropipette into the tube and touch the inside wall of the tube for better expulsion of the liquid.
3. Depress the plunger gently with the thumb to the first stop and go on till the second stop till all the fluid has been expelled.
4. Keep the plunger of the tip depressed and remove the pipette from the tube. Eject the tip by pressing the lever on the front.

Safe handling of microorganisms

Research in molecular biology involves the use of microorganisms. The research worker needs to strictly adhere to safe practices to avoid laboratory-acquired infection. The following precautions should essentially be taken:

1. A laboratory coat must be worn at all times. Gloves must be worn at all times with frequent changes.
2. Eating, drinking or smoking should not be allowed in the laboratory and applying make-up or licking gummed labels etc may be avoided.
3. Pipetting by mouth should be avoided.
4. All contaminated materials should be immediately placed in a disinfectant before disposal.
5. Spilling of contaminated stuff should be avoided. Books or other personal belongings should not be touched when work is in progress.
6. Any accident should be immediately referred to a Doctor.
7. Laboratory coat should be packed for laundering before leaving the lab.
8. Tidy work areas should be maintained.

Storage of cultures

There are several methods of culture storage available. Lyophilization by freeze-drying of bacterial cultures can keep the cultures alive for more than 30 years. Ultra-cold storage at –70 °C or in liquid N_2 is an other method. But the most popular method is the storage of cultures as glycerol stocks. Bacterial cultures can be stored in 50% sterile glycerol at –20 °C for a number of months. Stab cultures containing rich agar can also be prepared which can be stored at room temperatures for several months. For routine short-term storage and use in the laboratory, cultures can be streaked on agar plates or on slants and stored at 4 °C (in the refrigerator). However, if any specific strains with unique constructs are being stored, then it is advisable to add the appropriate antibiotics to the agar medium to exclude the growth of other undesirable strains of bacteria.

Storage of DNA and handling of enzymes:

DNA can be stored in TE buffer (pH 8.0) or in sterile distilled water at -20°C. Frequent freezing and thawing of DNA should be avoided to prevent breaks and nicks.

Enzymes should be stored in their original packing in the freezer at –20 °C and should never be placed on the counter top except in an ice bucket for short 1-2 minute intervals when pipetting is to be carried out. The assay buffers are supplied along with enzymes by the manufacturers and should also be stored at -20°C. Sterile double distilled water should be used for dilution during the preparation of reaction mixtures (restriction endonucleases) or mastermixes (for PCR). Enzymes should be handled very carefully to avoid contamination. Separate, fresh sterile pipette tips should be used for each job.

Use and care of the pH meter:

The method of operation of a pH meter (Fig-2) differs with different models. However, some general guidelines are given below:

Fig: 2: pH meter

1. The temperature compensation knob should be adjusted to the temperature of the test solution (usually the room temperature).

2. For calibration, the electrode should be dipped into a standard buffer (whose pH is accurately known) and the value confirmed on the dial. If the value is not correct, an adjustment can be made with the calibration knob.

3. Then the electrode is removed, washed well in distilled water, immersed in the unknown solution and the pH value read on the dial.

4. If a stir bar is in the beaker holding the unknown solution, its pH value can then be increased or decreased by adding drops of 1N NaOH or 1N HCl respectively, with continuous reading of the pH.

5. The electrode should be washed with distilled water and returned to the buffer.

Centrifuge and centrifugation

Centrifugation is a separation technique. Centrifugation depends on sedimention and gravitational forces (pull). For particles suspended in a specific liquid medium, the rate or velocity at which they sediment is proportional to the force applied.

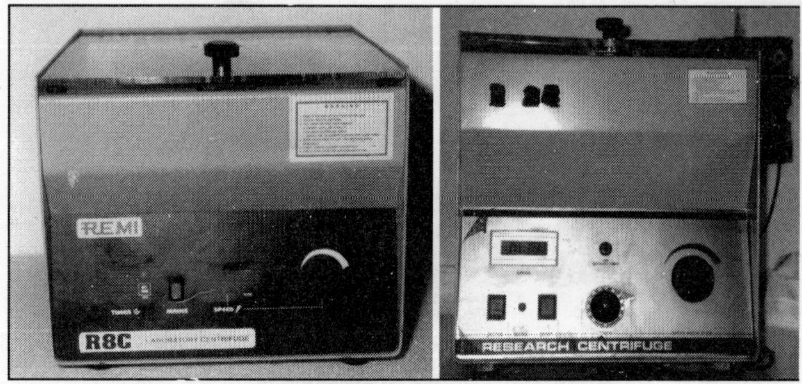

Fig: 3: Centrifuges

Thus the particles sediment more rapidly when the force applied is greater than the gravitational force of the earth. The gravitational pull under normal conditions is about 980 cm/sec or 1 g unit. If this force is increased, then very small particles will also be sedimented. This is provided by the centrifugal force generated by the revolutions of the rotor in a centrifuge (Fig - 3). Two basic types of rotors are the most common: fixed angle head and swinging bucket type rotors. Centrifuges which can produce speeds upto 20,000 rpm and more are generally refrigerated due to the high amount of heat produced by friction against the air. To achieve higher speeds, ultracentrifuges are operated with the rotor chamber under vacuum.

Use of a centrifuge

The centrifuge tubes are placed in their containers diametrically opposite to each other (if fewer tubes are used) and each pair should be equally balanced by weight. Care should be taken that the tubes are not filled to the brim. In case of the manually operated centrifuge, the speed must be increased gradually. After the centrifugiation is completed, the machine will gradually slow down until it stops completely. When conditions for the centrifugal separation of particles are reported, it is very essential that the rotor speed or revolutions per min (rpm) and duration of operation of rotor must be quoted with the gravitational forces (g), since the latter will vary depending on the size of the rotor.

METHODS IN MOLECULAR BIOLOGY

Spectrophotometry

Spectrophotometry often involves the measurement of a compound or group of compounds present in a complex mixture. Spectrophotometry and the related calorimetry measure light absorption. When light passes through a coloured solution, some wavelengths absorb more than the others do. Most compounds are able to absorb light in the visible region. Reaction with suitable reagents can render very minute quantities of material in the solutions to be measured. Generally the concentration of the compound being measured is proportional to the depth of the color. When a ray of monochromatic light of initial intensity Io passes through a solution in a transparent container, some of the light is absorbed, so that the intensity of the transmitted light I is less than Io. The Beer-Lambert Law explains this.

Beer's Law

When a ray of monochromatic light passes through an absorbing medium, its intensity decreases exponentially with increase in concentration of the absorbing medium.

Lambert's Law

When a ray of monochromatic light passes through an absorbing medium, its intensity decreases exponentially with increase in the depth of the absorbing medium.

If the depth of the solution is l, and l is kept constant then a plot of extinction (absorbance or optical density O.D) against concentration gives a straight line.

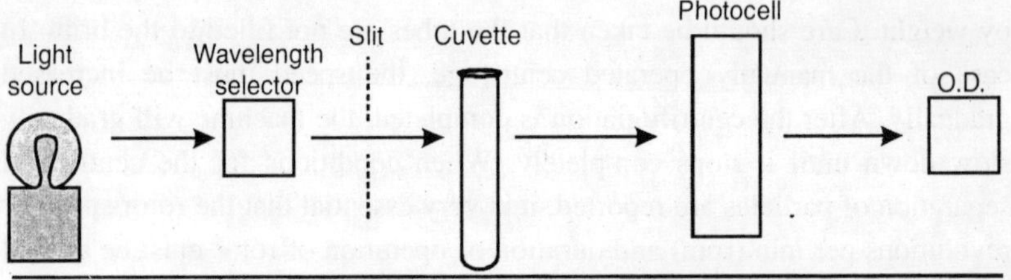

Fig: 4: Components of a spectrophotometer

In a spectrophotometer (Fig-4), a grating or prism provides monochromatic light and it is more sophisticated than the calorimeter in that it can distinguish between two compounds of closely related absorption strength. Compounds like nucleic acids absorb strongly in the ultra-violet region. Therefore an ultraviolet spectrophotometer (at 240 nm to 380 nm) is essential to measure the absorbance of DNA and RNA. This is explained in a later chapter.

***Use and care of spectrophotometer involves the following*:**

1. Cuvettes (containers used for measuring optical density) should be cleaned with 50% v/v nitric acid and rinsed with distilled water, before and after use.

2. The outside surface of the cuvette is wiped with a tissue before placing it in the holder.

3. Glass cuvettes can be used from 360 nm and above but for lower wavelengths, silica (quartz) cells are used.

4. The extinction (O.D) is adjusted to zero with a blank (containing all the reagents except the compound to be measured) and later the test solutions are measured.

5. Generally replicates should be used for accuracy.
6. A standard curve is prepared by using different known concentrations of the test compound and the concentration of the unknown compound is deduced from it.

Chromatography

Chromatography is one of the oldest methods of biochemical analysis. It involves the separation of components of a mixture by differential migration. Basically all chromatographic systems consist of the **stationary** and mobile phases. The stationery phase may be a solid, gel, liquid or a solid/liquid mixture that is immobilised. The mobile phase, which may be liquid or gaseous flows over or through the stationery phase. The choice of the stationary and mobile phases is made in such a way that the compounds to be separated have different distribution coefficients. There are several methods of chromatography: Paper, thin layer, gas liquid and column chromatography.

Paper chromatography

The cellulose fibres of the paper support the stationary liquid phase. The mobile phase passes along the paper sheet either by gravity feed or capillary action. This is the oldest method. There are two methods of chromatography, the ascending and descending chromatography. The sample is applied to the paper 2.0 to 2.5 cm from the edge by means of a micropipette or syringe. The spot is allowed to dry before applying some more of the sample on the same spot. In both ascending and descending methods, the solvent is placed in the base of a sealed tank or glass jar (the paper is held upright in the ascending and hung down from the upper **tank** in the descending method). The sample spots are always kept above the solvent level. Two-dimensional chromatography can also be done after the separation of the first dimension is completed. Methods of detection are employed to identify the separated compounds on the chromatogram. The identification of the compounds are made on the basis of their RF value. Simple methods of chromatography are explained in a later chapter.

Thin-layer chromatography (TLC)

A thin layer of the stationery phase is prepared on a flat glass or plastic or foil plate. The mobile phase moves rapidly across the layer thereby transferring the analytes (components of the mixture being subjected to chromatography) over it according to their distribution coefficients.

The movement of the analyte is expressed as RF value (see paper chromatography). A 0.25 mm thick layer is made on a 20 cm square plate by passing the plate spreader over it with the slurry. The plate is dried and sample is applied. The application of the sample is similar to the paper chromatography. Developing solvent is placed in a trough (up to a depth of 1.5 cm) which is kept closed for one hour to saturate it with solvent vapour. The plate is kept vertical in the trough and just enough solvent is used so that the sample spot stands above the solvent level. The trough is closed with the lid. Separation of analytes occurs within 50 minutes.

Two-dimensional TLC can also be carried out. The analytes are detected by spraying with suitable reagents and heating at 110°C which causes spots to develop. Several other methods like autoradiography, fluorescent dyes etc are also used for detection. A simple experiment using TLC is explained in a later chapter.

Gas liquid choromatography (GLC)

The principle of differential adsorption can be used to separate gases and vapourizable substances as well. In GLC, an inert, granular solid base is coated with a liquid like paraffin oil or silicon oil (grease). This material is packed into a narrow coiled glass or steel column 1-3 m long and 2-4 mm internal diameter, through which an inert carrier gas (the mobile phase) such as nitrogen, helium or argon is passed. The column is maintained in an oven at an elevated temperature, which keeps the compounds to be separated in a vapour state. The basis of separation is the difference in the partition coefficients of the volatilized compounds between the liquid and gas phases as the compounds are carried by the carrier gas. The compounds pass through a detector before passing out and are amplified on a chart recorder, which records a peak for each analyte.

Compounds like fatty acids, whose methyl esters can be easily vapourized, can be analyzed by GLC. Care is taken to use just enough solvent so that the sample spot stands above the solvent level. The trough is closed with the lid. Separation of analytes occurs within 50 minutes. Two-dimensional TLC is also carried out. The analytes are detected by spraying with suitable reagents and heating at 110°C which causes spots to develop. Several other methods like autoradiography, fluorescent dyes etc are also used for detection. A simple experiment using TLC is explained in a later chapter.

Column Chromatography

Column chromatography (Fig-5) revolves around the principle of a solid stationery phase and a liquid mobile phase that runs through it carrying the test compound.which settles some where in the column in the form of a band. The further addition of mobile phase causes the band fraction to travel lower and get collected.

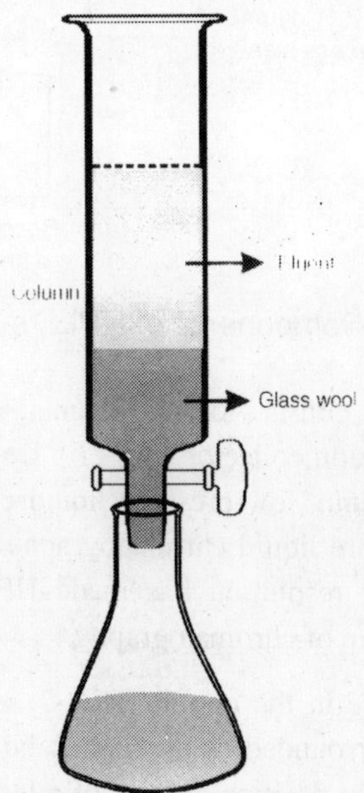

Fig 5: Column cromatography

The conventional solid phase is made by packing a slurry of 5 gm of alumina powder mixed with benzene. The column is tapped till the packing settles and more benzene is poured which acts as the mobile phase. The sample is extracted in benzene and traces of water are removed by solid anhydrous Na_2SO_4. This sample is dissolved in benzene and injected over the column.

In highly refined methods of column chromatography, the stationery phase attached to a suitable matrix (inert, insoluble support) is packed into a glass or metal column and the mobile phase is passed through the column either by gravity feed or by the use of a pumping system or applied gas pressure. A simple method of column chromatography is explained in a later chapter.

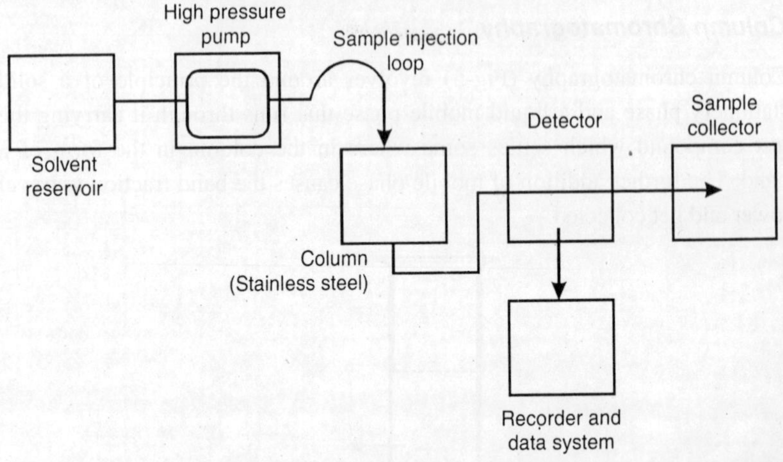

Fig: 6: Components of HPLC system

The chromatography unit consists of the column, a detector and recorder (computerized) and a fraction collector (Fig - 6). Depending on the pressure generated within the column, low-pressure liquid chromatography (LPLC) technique and high-pressure liquid chromatography (HPLC) techniques are in use. Faster and better resolution has made HPLC, the most popular, powerful and versatile form of chromatography.

The solid stationery phase in the column (Fig-7) is generally packed to a height of 5 cm. This is surrounded by the mobile liquid phase up to a height of 1 cm^3 labeled as the (A) position. 32 mg of a test compound is added to the column into the 1 cm^3 of mobile phase and it occupies the (A) position by displacing 1cm3 of mobile phase, which passes out from the base of the column. Another 1cm3 of the mobile phase is introduced (over the test compound) into the column and thus causes 16 mg of the compound to move down to (B) leaving 16 mg in (A). Then the compound gets redistributed so that 8 mg from (A) moves to (C). A further addition of 1 cm^3 and another 1 cm^3 continues the movement till stage V when a neat band is formed (A to E).

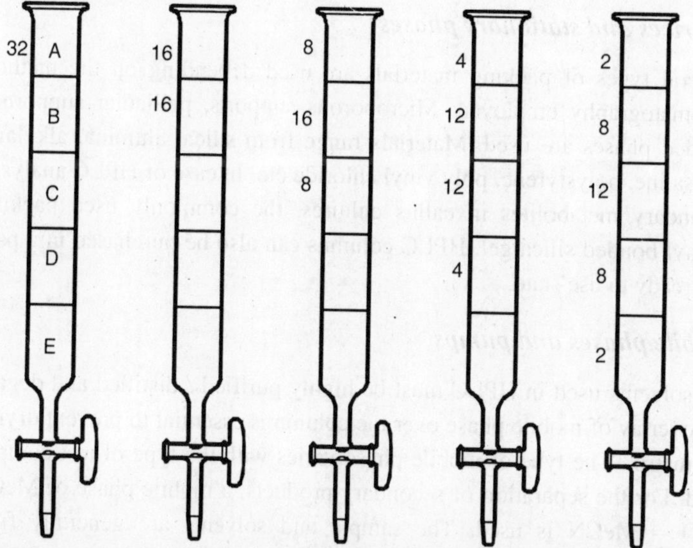

Fig: 7: Separation of compounds by column chromatography.

(32 mg of test compound is introduced which gets distributed into the "A", "B", "C", "D", and "E" areas. Mobile phase is added into "A" continuously for systematic movement.)

Every compound has a particular distribution coefficient. If two compounds with different distribution coefficients are added, two bands are formed. The separated compounds from the test compound are called analytes and they emerge in the effluent from the base of the column. If two compounds with different distribution coefficients are added, two bands are formed. The separated compounds from the test compound are called analytes and they emerge in the effluent from the base of the column.

Columns used in HPLC and LPLC

The columns used for HPLC are generally made of stainless steel and are manufactured to withstand very high pressure. Columns are generally 15 - 50 cm long with 1 - 4 mm diameter although microbore columns with 1 - 2 mm diameter but longer size are also used. Porous plugs of stainless steel or Teflon are used at both ends to retain the packing material.

Matrices and stationary phases

Several types of packing materials are used depending on the method of chromatography employed. Microporous supports, pellicular supports and bonded phases are used. Materials range from silica, alumina, alkylamine, octasaline, polystyrene, poly vinyl chloride etc. In case of HPLC analysis for secondary metabolites in callus cultures, the commonly used packing is phenyl bonded silica gel. HPLC columns can also be purchased in a packed and ready to use state.

Mobile phases and pumps

All solvents used in HPLC must be highly purified / distilled and degassed. An overlay of mobile phase over the column is essential to prevent drying of the matrix. The type of mobile phase varies with the type of test compound used. For the separation of secondary products, a mobile phase of MeOH -- H2O -- MeCN is used. The sample and solvents are generally filtered through 0.45 mm filter paper.

Sample preparation and application

The extraction and purification of test compounds is a multi stage process. A solvent extraction method is used for extraction of organic compounds from aqueous mixtures by diethyl ether or dichloro methane. If there is any trace of water it is extracted by an anhydrous salt such as sodium sulphate. The sample (after drying) is dissolved in methanol for application. Solid phase extraction by the use of disposable columns is also used. The sample in methanol can be filtered through 0.45 mm filters. The sample is applied (about 10 m1) onto the column or into an inert plug on top of it with a micro-syringe.

Column development and sample elution

The components of the applied sample are separated by the continuous passage of the mobile phase through the column. This leads to elution development. High-pressure pumps are used which maintain a constant uniform flow.

Detectors and fraction collection

As the analytes emerge in the effluent from the column, it is necessary to detect their presence. Coloured analytes can simply be observed visually but colourless compounds are identified by U V absorption or fluorescence spectroscopy. The analytes can be collected for further confirmation or analysis such as NMR.

Plotting of the chromatogram

The chromatograms are plotted with the absorption maxima of the analyte at a particular wavelength. Each peak formed refers to a particular analyte. The time taken for each analyte peak to emerge from the column is referred to as its retention time. This is specific to a particular compound.

Quantification and standards

After the chart recording is obtained, the area of each peak is determined by measuring the height of the peak (*hp*) and its width at half the height (*wh*). The product of *hp* and wh gives the area of the peak. The retention time and peak areas are calculated (both values are similar).

The amount of analyte present may be determined by the use of a calibration curve obtained by using standard analytes. These can be used as internal standards (when added to the sample at a very early stage) or as external standards (added at a very late stage) for accurate quantification.

In normal phase liquid chromatography, the stationary phase is a polar compound such as an alkyl nitrile or an alkylamine and the mobile phase is a non-polar solvent such as hexane. For reversed-phase liquid chromatography, the stationary phase is a non-polar compound such as octasaline and the mobile phase is a polar solvent such as methanol. The reversed-phase HPLC is the most widely used form of chromatography mainly because of its flexibility and high resolution. It is widely used to analyze drugs, metabolites, pesticide residues and amino acids. The technique applied to proteins is referred to as FPLC (fast protein liquid chromatography).

Electrophoresis

Electrophoresis refers to the migration of charged particles under the influence of a direct electric current. Many important biological molecules such as amino acids, peptides, proteins, nucleotides and nucleic acids possess ionizable groups and therefore at a given pH, exist in solution as electrically charged species either as cations (+) or anions (-). The particles with a net charge will migrate to the cathode or anode. Electrophoresis systems consist of a power pack to supply the electric current and the unit in which the separation of the molecules is accomplished. The units are available for vertical (Fig-8) or horizontal (Fig-9) slab gel systems. However, the earlier systems were based on filter paper or cellulose acetate strips or silica or alumina strips. The earliest gel system was the starch gel, which is still in use. Polyacrylamide gel is a recent support media for electrophoresis and it has been in use for tube gels for a number of years. However, the most modern system for electrophoresis is the slab gel. For separation of proteins, SDS-PAGE (sodium dodecyl sulphate - polyacrylamide gel electrophoresis) is preferred for its clear resolution and capacity to form distinct bands besides being suitable for molecular weight determination of proteins. The pores of polyacrylamide gel are formed by the cross-linking of acrylamide with N, N methylene bis acrylamide through a controlled polymerization. Acrylamide gels are also used to separate small DNA fragments. Gels of low percentage (e.g. 8 %), that have large pore sizes are used to separate DNA and gels of higher percentage (e.g. 10 - 30 %) are used for separation of proteins due to their smaller pore size. Vertical slab gels are used routinely for the analysis of proteins and for the separation of DNA fragments for DNA sequence analysis. Electrophoretic methods are explained in detail in later chapters.

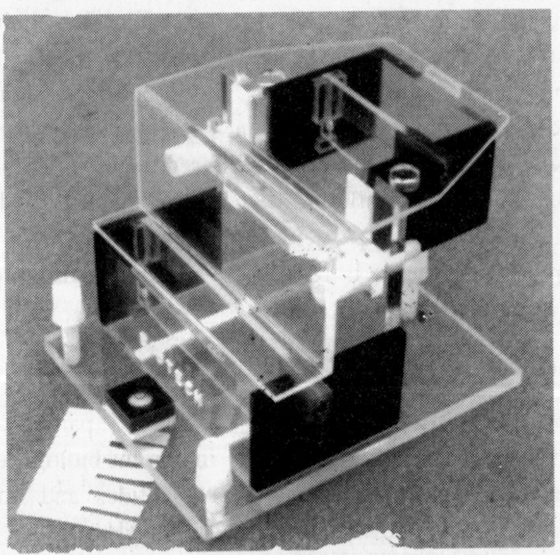

Fig. 8: Vertical electrophoresis unit

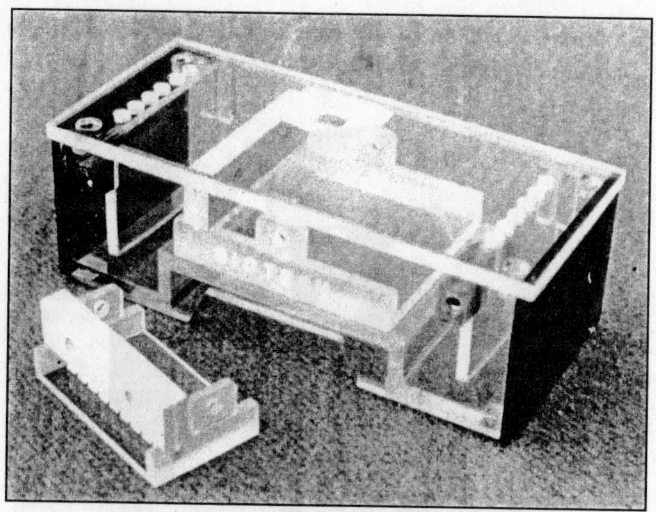

Fig. 9: Horizontal electrophoresis unit

Another matrix, agarose is a preferred gel substrate for many purposes. Large pore size gels (of low concentration) are preferred for separation of DNA and RNA fragments. The gelling properties of agarose are stimulated by its solubilization through heating, with inter and intra molecular hydrogen bonding within the molecules occurring as the agarose cools. The bonding also results in pores. Agarose gels are mostly used in horizontal slab systems. Due to their poor elasticity they are not preferred for vertical slabs. Agaroses of different melting points are available, and the type with a low melting point is useful for eluting DNA bands from gels.

Microscopy

Cell, tissue or organelle preparations often need to be evaluated for their integrity of structure. Microscopy fulfills this need. Microscopy involves magnification and improved resolution (the ability to visualize two closely placed objects as separate entities). Light microscopes (Fig -10) operate with optical lenses to sequentially focus the image of objects whereas electron microscopes work with electromagnetic lenses, in a transmission or scanning mode. Polarized light microscopes are used to detect optically active substances in cells (e.g. starch granules in amyloplasts). Phase-contrast light microscopes improve the image contrast of unstained materials. The confocal microscope uses a laser beam to visualize images, enabling relatively thin sections to be examined at several focal planes, with subsequent digital assembly of a composite image.

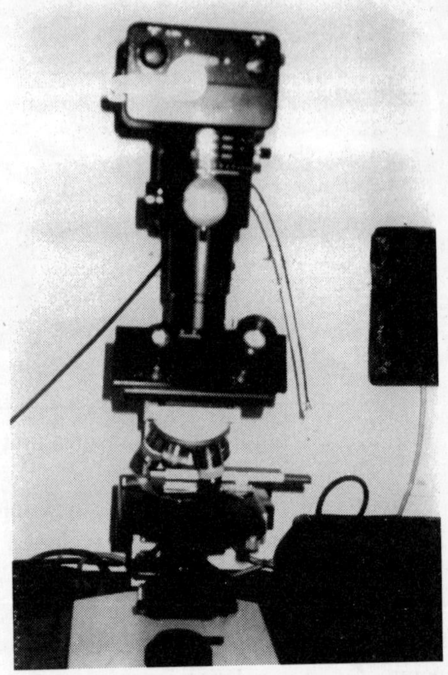

Fig: 10: Microscope with camera

Light microscopes magnify approximately 1500 times and have a resolution limit of about 0.2 mm whereas a transmission electron microscope (TEM) is capable of magnifying approximately 200, 000 times and has a resolution limit of 1 nm. Scanning electron microscope (SEM) uses a fine beam of electrons to scan back and forth. The resolution limit of a SEM is 6 nm. Several other sophisticated methods of microscopy are being developed like the ion-probe analysis through modifications of election microscopy procedures.

Microscopy requires relatively thin samples of specimens (i.e. squashes, smears or sections). Preservation of tissue is accomplished by chemical fixation using chemicals like acetic acid and alcohol. Fixation for electron microscopy is carried out in formaldehyde, glutaraldehyde or osmium tetroxide. Ultra - rapid freeze fixation is also utilized (in liquid propane or nitrogen) and has advantages for preservation of fine structure. Sections of frozen tissue are prepared with an ultra-cryomicrotome. For the TEM, heavy metal salts are used which have selective affinity to cell organelles. Negative staining (phospho-tungstic acid) and freeze fracturing techniques are also used. The images are recorded by photomicrography. For light microscopy, histological stains produce the contrast for better resolution. Acidic components bind successfully to basic stains (like aceto-orcein and aceto-carmine).

Micrometry

Micrometry is a method of measurement of micro organisms or microscopic objects like cells or tissues which are expressed in micro meters (mm) which is a thousandth of a millimeter (mm).

The measurement involves the use of an ocular micrometer placed inside the eyepiece. The graduations on the ocular are calibrated against the scale of measurement on a stage micrometer kept on the stage of a microscope. The stage micrometer is a glass slide with one mm etched on it in 100 divisions. Since one mm is equivalent to 1000 mm, each stage division (i.e. each of the 100 divisions) is equivalent to 10 mm.

The ocular divisions are super imposed over those of the stage divisions by viewing through the eyepiece and rotating it. The stage is moved to facilitate the viewing and the microscope adjusted for maximum clarity. The number of ocular divisions included between two stage divisions are counted from the left and the calibration of the ocular is done by the following calculation.

$$\text{One ocular division (in mm)} = \frac{\text{Number of divisions on stage}}{\text{Number of divisions on ocular}} \text{X} \, 10$$

(Calibration factor)

Example:

If 6 ocular divisions are included between 2 stage divisions,

$$\text{then, one ocular division (1 mm)} = \frac{20 \text{X} 10}{6} = 3.3$$

Two to three readings are recorded and the average value is taken for calibration.

Consider that the ocular is used to measure the length of a chloroplast.

If the cell measures 3 ocular divisions, it means that the length of the cell is 3 x 3.3 = 9.9 mm., where 3.3 is the calibration factor.

Radioactivity and autoradiography

Radioactivity

The protons, electrons and neutrons make up the predominant particles present in an atom. Protons and neutrons form the nucleus of the atom and electrons revolve around it in orbitals. The number of protons in an atom make up the atomic number. Electrical neutrality is achieved by the presence of equal number of electrons and protons. The total weight of neutrons and protons in an atom make up its atomic weight.

It is possible to have different number of neutrons in different atoms of the same element although the number of protons remains the same. Such atoms are called isotopes. Some types of isotopes undergo decay and release some energy. These are called radioisotopes. The energy is released in the form of beta and gamma radiations. Due to radiation, the basic structure of an atom is altered and it becomes a different element. This is called transmutation. The decay of a radioisotope can be detected by measuring the emitted radiation by suitable methods and it is an irreversible process. The time required for the radioactivity in the sample to be reduced to half of its original level is called its half-life.

There are three methods available to measure radioactivity or radiation emitted by a radioactive compound:

1. ***Autoradiography*** *wherein* the radiation exposes the X-ray film.
2. ***Ionization*** wherein the collision of the particles with other particles cause ionization and is measured by a Geiger Muller tube.
3. ***Scintillation counting*** where the collisions with some substances like phosphorus emit light flashes which can be counted by the liquid scintillation counter. The radioactivity is measured in terms of counts per minute (CPM) and the activity expressed as Micro Curies (μCi) or disintegrations per minute (DPM).

$$DPM = CPM / \text{Efficiency for the isotope used.}$$

$$\mu Ci = DPM / 2.22 \times 10^6.$$

Autoradiography

Autoradiography is a technique used in situations where the distribution of radioactivity in a biological sample or a chromatogram is to be determined. Autoradiography is used to study the movement of solute in whole plants or tissues and selective synthesis of macro molecular components like DNA at various sites in the cell.

The sample containing radioactivity (i.e. a filter or a chromatogram with labelled compounds or a tissue slice with cells harbouring radioactive compounds) is kept in close contact with an ordinary X - ray film. The emanating radiations affect the chemicals in the photographic plate in the same way that light exposes the photographic film resulting in the deposition of silver grains.

After suitable time of exposure, the X - ray film is developed. The areas that were exposed to radioactive emission appear, as darkened regions while unexposed regions will be clear. A densitometer can be used to measure the intensity of the spot results from the exposure, however, since the film reaches a plateau of darkness, quantification from an X - ray film is only accurate over a narrow range. Some of these procedures are explained in a later chapter.

> *Safety precautions have to be adhered to while handling radioisotopes and all manipulations should be carried out with gloved hands and behind a plexi-glass hood. Proper disposal of the radioactive wastes should be carried out. A thorough check of the work area with a Geiger counter should be carried out after each session of radioactive work.*

Isolation of nucleic acids

To protect the integrity of nucleic acids, it is wise to carry out the isolation procedure at 4°C. DNA can be protected by autoclaving all solutions and glassware and by the use of EDTA to chelate magnesium ions needed for DNase activity. Subsequent steps are regulated according to the need. Cells and cell membranes are disrupted by the use of a detergent or by grinding in liquid nitrogen. For the isolation of DNA, ribonuclease treatment is carried out to remove RNA. Water saturated phenol or phenol-chloroform mixture is added and proteins removed, before the DNA is precipitated in ehtanol. Some care must be taken to isolate high molecular DNA: i.e. the DNA -

containing solution should not be vortexed and the procedures should not use glass vessels or pipettes which will cause shearing of DNA. Because RNAs are not as large molecules as DNA, the RNA isolation procedure does not need to be as gentle and rapid stirring can be used. All glassware should be cleaned with detergent or other special chemical mixtures and the fingers should not be in contact (gloves to be worn) with the material to minimize the exposure to RNases. Deproteinization is followed by the DNase treatment before RNA can be precipitated.

Electrophoresis to separate DNA molecules

Most widely used gels are the horizontal agarose gels, though polyacrylamide gels are also used. Electrophoresis is followed by staining with ethidium bromide. Restriction endonucleases are used to cut the DNA and the restriction fragments electrophoresed resulting in their separation according to their sizes. The piece of gel containing the desired DNA fragment can be cut out and the DNA recovered from it by elution. This is explained in detail in later chapters.

Southern and Northern hybridization

Southern hybridization is named after its inventor E.M. Southern, and is used to identify DNA that has homology to a probe such as a transgene. In this procedure (Fig-11), the genomic DNA is isolated, digested by restriction endonucleases and the restrictions fragments separated by agarose gel electrophoresis followed by a transfer of the of the bands on the gel to a membrane by the Southern blot method where large volumes of buffer are drawn through the gel and through the membrane by capillary transfer. The bands transferred (southern blot/transfer) onto the nitro cellulose paper can then be denatured and treated with a short radioactive DNA probe which is specific to the DNA insert in question. The probe hybridizes with the specific insert at particular bands indicating the position of the DNA insert.

Northern blotting is a procedure used to characterize RNA molecules. Electrophoresis of the (RNA) fragments is followed by blotting onto a nitrocellulose filter in the same way as the Southern blot. Proteins can also be visualized by electrophoresis, blotting onto nitrocellulose membranes and probed by antibodies (known as the Western blot). These methods are explained in a later chapter.

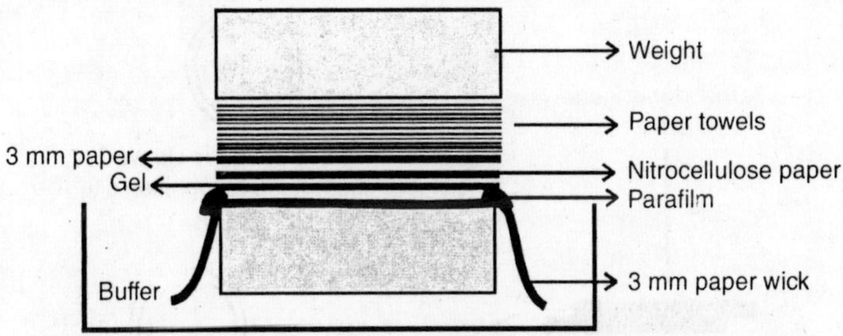

Fig. 11: Set up of Southern blotting

DNA sequencing

DNA sequencing methods have resulted in a better understanding of the gene structure. Two methods are in use. The di-deoxy nucleotide method of Sanger and the chain termination method of Maxam and Gilbert. Both the methods are based on the electrophoresis of DNA.

Cloning and genetic engineering

Cloning describes the procedures (Fig - 12) through which segments of DNA are identified, isolated and inserted in vectors (recombinant plasmids or constructs). These are then transformed into host bacterial cells to be multiplied or stored for future use. Constructs that carry a gene of interest are then isolated and transferred into plant cells in a process called genetic engineering. The transformed plant cells may be cultured *in vitro* to regenerate transgenic plantlets. The presence of the foreign gene in these putative transgenic plants can be confirmed through various molecular methods including PCR, Southern blot hybridization etc. Some of these methods are explained in later chapters.

Molecular biology in association with genetic engineering has opened up several exciting possibilities. It has been possible to clone genes for human insulin, growth hormone, interferons, tumour necrosis factor, blood clotting factor VIII and viral coat proteins (vaccines) in bacteria so that they are expressed and their gene products (poly peptides) can be recovered from cell cultures. Altered enzymes can also be produced by tailoring genes according to need.

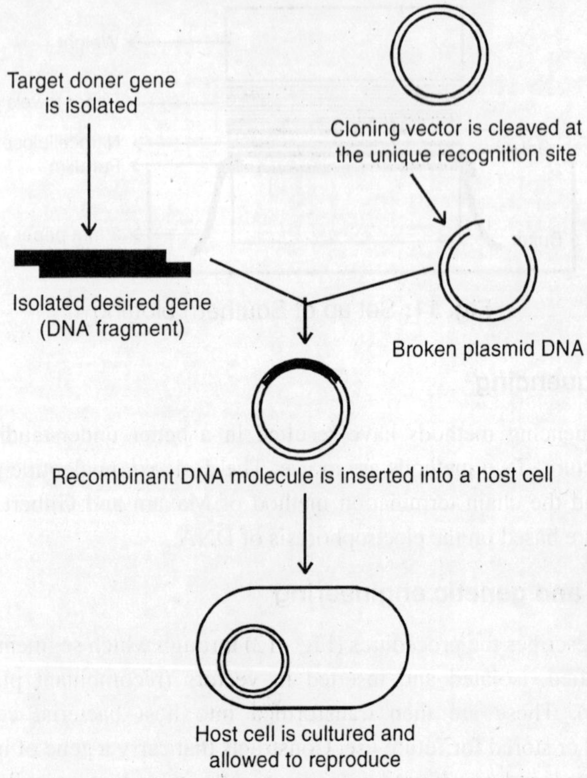

Fig. 12: Gene cloning in bacteria.

Polymerase chain reaction

Polymerase chain reaction is one of the most powerful techniques of molecular biology. It is used for amplification of specific segments of DNA. If the sequences of the flanking regions of a DNA molecule are known, the unknown region can be amplified to generate a large quantity of DNA. Primers are constructed for the flanking DNA sequences and used to synthesize the target sequence with a heat tolerant DNA polymerase (usually, Taq DNA polymerase) and addition of sufficient dNTPs.

The DNA has to be denatured initially so that the primers can anneal and DNA synthesis can initiate. This requires a high temperature of 950C. The *E. coli* DNA polymerase was thermolabile and could not withstand high temperatures required during the denaturation process of DNA. This problem was solved by the use of the enzyme extracted from a thermophilic bacterium, *Thermus aquaticus.* The enzyme Taq DNA polymerase withstands high temperature. It is used in the thermocycler which gets heated up to denature the DNA and regulates the production of the strands for every cycle).

Fig. 13: Thermocycler

The thermocycler or the PCR machine (Fig-13) regulates these cycles of denaturing, annealing and synthesis completing each cycle within 4 minutes. A thoroughly clean environment has to be maintained to avoid contamination with foreign DNA. PCR has become an indispensable tool in many major forensic science laboratories since gene amplification is possible even from a tiny quantity of DNA (hair or a drop of blood or semen). Experiments dealing with the PCR method are dealt with in a later chapter.

REFERENCES

1. Brown, S. B. (1980) An introduction to spectroscopy for biochemists. Academic Press, New York.
2. Gueffrey. D. E. (1986) Buffers: A guide for the preparation and use of buffers in biological systems. Calbiochem Biochemicals. San Diego.
3. Hames. B. D. and Rickwood. D. (1987) Gel electrophoresis of proteins: A practical approach. IRL press. Oxford.
4. Harris, D. A. and Bashford, C. L. (1987) Spectrophotometry and spectrofluorimetry: A practical approach. IRL press. Washington D.C.
5. Segal. I. H. (1976) Biochemical calculations. 2nd Edition. John Wiley and Sons. New York.

2

Plant Biochemistry and Physiology

The experiments included in this chapter have been designed for a broad spectrum of interest areas in the use of biochemical methods in the plant sciences. Biochemistry can be applied for the improvement of goods and services in a wide variety of industries like food, drinks, pharmaceuticals, environmental protection, biotechnology and the health service. Routine biochemical analysis forms an essential part of the quality control aspects in many industries and special biochemistry departments are generally set up for this purpose. From another perspective, comparative plant biochemistry can yield valuable basic information about the integration of biochemical pathways and products.

The experiments in this chapter include qualitative tests and quantitative estimations besides enzyme activity studies. The range includes carbohydrates, amino acids, proteins, lipids, nucleic acids, chloroplast pigments apart from several diverse physiological estimations like total and titratable acidity, ascorbic acid content, IAA content, water potential besides the activities of several enzymes.

SAMPLE PREPARATION

For the qualitative and most of the quantitative tests described in this chapter, "test samples" can be prepared with the commercially available pure chemicals. The concentration of these samples is however specified in each case. Appropriate negative controls are included for comparison. In quantitative estimations from plant materials, positive controls comprising the pure chemicals are necessarily maintained.

QUALITATIVE TESTS

The qualitative tests for carbohydrates and amino acids are described here.

Qualitative tests for carbohydrates

General tests for presence of carbohydrates

Molisch's test

Concentrated sulphuric acid hydrolyzes glycosidic bonds to give monosaccharides, which are then dehydrated to furfural and its derivatives. These products then combine with sulphonated α - napthol to give a purple compound.

- *Test solution*: Prepare a 1% solution of any carbohydrate (e.g. sucrose). Use distilled water as the negative control.
- *Test:* Add a few drops of α - napthol to 1 ml of test solution and also to the negative control. Then carefully add 1 - 2 ml of concentrated H2SO4 down the side of the test tube to form two layers. A red-violet ring forms between the two layers, indicating the presence of carbohydrates.

Anthrone's reaction

The principle is the same as above but the furfural reaction with anthrone gives a blue-green complex. Anthrone solution can be prepared by dissolving 200 mg anthrone in 100 ml of concentrated H_2SO_4.

- *Test solution:* Prepare a 1% solution of any carbohydrate (e.g. glucose). Use distilled water as the negative control.
- *Test:* Add 2 drops of test solution to 2 ml of anthrone. Carry out the same with the negative control. A blue-green colour complex develops, indicating the presence of carbohydrates.

Reactions for presence of reducing sugars

Fehling's test

Fehling's solution is prepared just before use by mixing equal volumes of 7% $CuSO_4$ and alkaline Na.K tartarate (Dissolve 24 g KOH and 34.6 g of sodium potassium tartarate in 100 ml dH_2O).

- *Test solution:* Prepare a 1% solution of any monosaccharide (e.g. glucose). Use distilled water as the negative control.
- *Test:* Add 5 ml of Fehling's solution to a few drops of test solution and boil in water bath for a few minutes. Reddish brown colour is obtained, indicating the presence of reducing sugars.

Benedict's test

Benedict's solution is a modified preparation of Fehling's solution. Dissolve 4.3g of finely pulverized $CuSO_4$ in 25 ml of hot water, cool and dilute to 40 ml with water. Dissolve separately 43g of sodium citrate and 25g of anhydrous sodium carbonate (or equivalent of hydrate) in 150 ml of water. Heat to dissolve. Cool. Add $CuSO_4$ solution and dilute to 250 ml. Keep in a cork stoppered bottle.

- *Test solution:* Prepare a 1% solution of any monosaccharide (e.g. glucose). Use distilled water as the negative control.
- *Test:* Add 5 drops of test solution to 2 ml of Benedict's reagent in a water bath. Rust brown colour is obtained, indicating the presence of reducing sugars.

Barfoed's test

Barfoed's reagent is weakly acidic and is reduced by mono and disaccharides. The reagent is prepared by dissolving 13.3 g of copper acetate in 200 ml water and adding 1.8 ml of glacial acetic acid to it.

- *Test solution:* Prepare a 1% solution of any mono or disaccharide. Use distilled water as the negative control.
- *Test:* Add a few drops of test solution to 2 ml of Barfoed's reagent, boil in water bath and allow to stand at room temperature. Brick red colour is obtained, indicating the presence of monosaccharides.

Bial's test

This is a test for pentoses. When pentoses are heated with concentrated HCl, furfural is formed which reacts with orcinol in the presence of ferric ion to give a blue-green colour. The reaction is not absolutely specific for pentoses since longer heating of some hexoses yields hydroxy methyl furfural which also reacts with orcinol to give a green complex. Prepare the reagent by dissolving 150 mg orcinol in 50 ml of concentrated HCl.

- *Test solution:* Prepare a 1% solution of any pentose sugar. Use distilled water as the negative control.
- *Test:* Add a few drops of test solution to 3 ml of reagent and heat just to a boil. A blue-green colour is obtained as proof of pentoses.

Seliwanoff's test

This test is for ketoses. Ketoses are dehydrated more rapidly than aldoses to give furfural derivatives, which reacts with resorcinol to give a red complex. Prepare resorcinol reagent by dissolving 50 mg resorcinol in 100 ml 3NHCl.

- *Test solution:* Prepare a 1% solution of any ketose sugar (e.g. fructose). Use distilled water as the negative control.
- *Test:* Add a few drops of test solution to 2 ml of resorcinol reagent and boil for 1 min. Note the appearance of deep red colour.

Test for sucrose

Sucrose is a common non-reducing disaccharide and it does not reduce alkaline CuSO4 solution. Sucrose is therefore hydrolyzed in acid solution to give glucose and fructose, which are then tested.

- *Test solution:* Prepare a 1% solution of sucrose. Use distilled water as the negative control.

- *Test:* Add 5 drops of concentrated HCl to a few drops of sucrose solution. Heat for 5 to 10 min in boiling water bath, cool and add a few drops of 1 N NaOH to neutralize it. Then carry out the reduction test and Seliwanoff's tests.

Test for starch

Iodine test

Iodine forms complexes with polysaccharides. Starch gives a blue colour with iodine solution while glycogen and partially hydrolyzed polysaccharide gives a red colour.

- *Test solution:* Prepare a 1% starch solution and heat it till the starch dissolves completely. Use distilled water as the negative control.
- *Test :*Acidify the test solution with dilute HCl (0.1 N HCl). Then add 2 drops of iodine solution (0.005N iodine solution in 3% potassium iodine) and observe the colour.

QUALITATIVE TESTS FOR AMINO ACIDS

General tests

Ninhydrin reaction

Ninhydrin is an oxidizing agent (reagent) and between pH 4 to 8, it reacts with all α - amino acids to give a purple coloured compound. The reaction is also given by primary amino acids and ammonia but without the liberation of CO_2. Imino acids (proline and hydroxy proline) also react with ninhydrin but in this case, a yellow colour is obtained instead of the usual purple colour.

- *Test solution:* Prepare a 1% solution of any amino acid. Adjust the pH to almost neutrality. Use distilled water as the negative control.
- *Test:* Place 1 ml of amino acid solution in a test tube. Add 5 drops of concentrated ninhydrin and boil for few minutes. Appearance of blue colour indicates the presence α-amino acids.

Reactions for sulphur amino acids

Lead sulphide test

The lead sulphide test is to detect cysteine (cys-SH) and cystine (cys). When the cys-SH and cys are heated in strong alkali (NaOH). Some of the sulphur is transformed to sodium sulphide, which can be detected by precipitation as lead sulphide from alkaline solution. The sulphur of methionine is not affected by this reaction. The reagent is prepared by dissolving 45g NaOH in 240 ml distilled water and then dissolving 7.5 g litharge (PbO) in the hot caustic solution.

- ❑ *Test solution:* Prepare 1% solutions of cysteine and cystine. Use distilled water as negative controls.
- ❑ *Test* :Boil 2 ml of the test solution with 1 ml of the reagent. Appearance of a black precipitate indicates the presence of cysteine or cystine.

Nitro prusside test

Test for cysteine

The thiol group of cysteine reacts with sodium nitroprusside in the presence of excess ammonia to give a red colour.

- ❑ *Test solution:* Prepare 1% solution of cysteine. Use distilled water as the negative control.
- ❑ *Test:* Mix 0.5 ml of freshly prepared 10% sodium nitroprusside with 2 ml of test solution. Boil for 2 - 3 min and add a drop of strong ammonia. Appearance of deep red colour indicates cysteine.

Test for cystine

Repeat the above test with cystine after mixing equal volumes of 1% cystine and 1%NaCN or KCN (**caution ! Use gloves**. CN is extremely poisonous) and leave it at room temperature for a few minutes for colour development.

Reactions of aromatic amino acids

Xantho proteic reaction

Amino acids that contain aromatic rings form yellow derivatives (nitro derivatives) upon heating with concentrated HNO3. The salts of these derivatives are orange in colour.

- *Test solution:* Prepare a 1% solution of any aromatic amino acid (e.g. tryptophan). Use distilled water as the negative control.
- *Test:* Add an equal volume of concentrated HNO_3 to 1 ml of amino acid solution. Cool and observe the colour change. Add sufficient 40% NaOH to make the solution strongly alkaline. A yellow colour in acid solution, which turns a bright orange with alkaline, constitutes a positive result. Phenylalanine gives a weakly positive reaction since no ortho directing group is present.

Millon's reaction

This reaction is basically for tyrosine. Compounds containing the hydroxy benzene radical react with Millon's reagent to form red complexes. The original Millon's reagent was a solution of mercuric nitrate in 50% HNO_3. But modifications are now available that are less subject to interference from inorganic salts.

- *Test solution:* Prepare a 1% solution of tyrosine. Use distilled water as the negative control.
- *Test :* Add 5 drops of Millon's reagent to 1 ml of test solution and keep in a boiling water bath for 10 minutes. Cool to room temperature and add 5 drops of 1% sodium nitrite solution. A positive result is indicated by a brick-red colour.

Glyoxylic test

Glyoxylic test is basically for tryptophan. The indole group of tryptophan reacts with glyoxylic acid in the presence of concentrated H_2SO_4 to give a purple colour. Glacial acetic acid that has been exposed to the light will have glyoxylic acid, so it can be used for the test.

- *Test solution:* Prepare a 1% solution of tryptophan. Use distilled water as the negative control.
- *Test:* Add 2 ml of glacial acetic acid to 1 ml of the test solution, and then pour about 2 ml of concentrated H_2SO_4 carefully down the sides of a sloping test tube to form two layers. Appearance of a purple ring at the liquid junction indicates a positive reaction.

Pauli's test

This is for tyrosine and histidine. Diazotized sulphonic acid reacts with amines, phenols and imidazoles to form highly reactive azo-compounds. The diazonium compound is formed only in the cold. So all the solutions are cooled in an ice bath before diazotization.

- ❑ *Test solution:* Prepare a 1% solution of tyrosine or histidine. Use distilled water as the negative control.
- ❑ *Test :* Mix 1 ml of concentrated sulphonic acid with 2 ml of test solution (cooled on ice) and 1 ml of 1% sodium nitrite solution and leave in cold for 3 minutes. Make the solution alkaline by the addition of a few drops of 1% Na_2CO_3. Appearance of a red -orange colour indicates a positive test for tyrosine and histidine.

Ehrlich's test

Ehrlich's test is for tryptophan. Ehrlich's reagent reacts with a number of organic compounds such as indoles and aromatic amines to give orange complexes. Ehrlich's reagent consists of paramethyl amino benzaldehyde.

- ❑ *Test solution:* Prepare a 1% solution of tryptophan. Use distilled water as the negative control.
- ❑ *Test:* Add 2 ml of Ehrlich's reagent to 0.5 ml of test solution and observe the change from red to orange colour.

Sakaguchi test

The Sakaguchi test is used for arginine, which is the only amino acid with a guanidine group. It reacts with α - napthol and an oxidizing agent such as bromine water to give a red colour.

- ❑ *Test solution:* Prepare 1% solution of arginine. Use distilled water as the negative control.
- ❑ *Test:* Mix 1 ml of strong alkali with 2 ml of amino acid solution and add 2 drops α -napthol. Shake well and add 4 - 5 drops of bromine water. Formation of red colour indicates arginine.

Test for proline and hydroxy proline

Test for proline and hydroxyproline is carried out by the Isatin reagent.

- *Test solution:* Prepare 1% solution of proline or hydroxyproline. Use distilled water as the negative control.

- *Test:* Take a drop of amino acid solution on a Whatman No.1 filter paper and dry the spot by exposing it to flame or heater. Then spray the paper with 1% Isatin solution and dry it again. Formation of blue spot indicates proline.

QUANTITATIVE ESTIMATIONS

Estimation of carbohydrates by anthrone method

Hydrolysis of carbohydrates with H_2SO_4 produces furfural, which reacts with anthrone to form a green coloured complex that can be measured calorimetrically or spectrophotometrically. Prepare 0.2% anthrone reagent in concentrated H_2SO_4.

Standard graph

Prepare standard glucose solution (10 mg/ml) and pipette out into a series of test tubes, different volumes of glucose solutions (containing 0 - 100 μg) and bring the volume of each to 1 ml with distilled water. Add 4 ml of ice cold concentrated anthrone reagent into each tube and mix well. Keep the tubes in boiling water for 10 min or till the green colour develops. Cool to room temperature and measure the optical density (O.D) at 620 nm using a blank without the glucose. Draw the standard graph as shown in Fig-14 by plotting the OD values on the Y-axis and the glucose concentration (μg/ μl) on the X-axis.

[The amount of glucose in the unknown sample can be read out through its O.D. value at 620 nm.]

Sample

Proceed as above using 0.5 ml or 1 ml of the unknown sample (take any concentration at random). Measure the O.D and read out the amount of glucose contained in the sample from the standard graph.

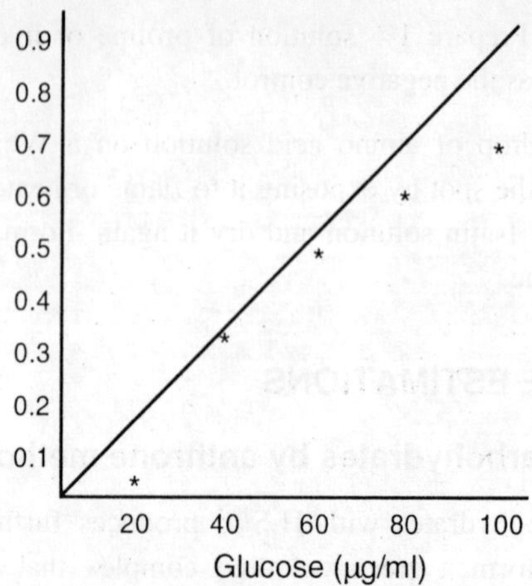

Fig 14: Glucose standard graph

Estimation of reducing sugars (glucose) by 3,5-dinitro salicylate method

The reagent 3,5-dinitro salicylic acid is reduced by glucose and a reddish colour is produced whose intensity can be measured spectrophotometrically. Prepare the reagent by dissolving 1 gm of 3,5-dinitro salicylic acid in 20 ml of 2N NaOH and 50 ml of distilled water. Then add 30 gm of Na.K. tartarate and bring up to 100 ml with distilled H_2O.

Standard graph

Prepare standard glucose solution (10 mg/10ml) and pipette out different volumes of glucose solutions (containing 100 µg - 1000 µg of glucose) into a series of test tubes and add distilled water to bring each sample to 1 ml. Add 2 ml of the 3,5-dinitro salicylic acid reagent, shake well and place in a boiling water bath for 5 min. Cool the tubes to room temperature. Add 7 ml of distilled water to each tube and measure the optical density (O.D) at 540 nm against a blank without the glucose. Prepare the standard graph.

Unknown sample

Proceed as above using 0.5 and 1 ml of the sample glucose solution (take any concentration at random), measure the O.D and read out the amount of glucose the sample contains from the standard graph.

Estimation of fructose by Roes resorcinol method

Combined and free keto hexoses get dehydrated to give furfural derivatives, which react with resorcinol to give a fairly red colour, which can be measured by a spectrophotometer. Prepare the resorcinol reagent by dissolving 0.1 gm resorcinol and 0.25 gm thiourea in 100 ml of glacial acetic acid. Prepare 5% HCl.

Standard graph

Prepare a standard fructose solution (10 mg/100 ml) and pipette out different volumes of fructose solutions (containing 20 - 80 μg) into a series of test tubes and add distilled water to make them up to 1 ml each. Add 1 ml of resorcinol reagent, followed by 8 ml of HCl solution. Then shake the tubes to mix well. Place the tubes in a water bath at 80°C for 10 min until the red colour develops. Then remove them and immerse in tap water until they cool to room temperature. Then, measure their optical densities at 550 nm within 30 min against a blank without the fructose. Prepare the standard graph.

Unknown sample

Proceed as above with 0.5 and 1 ml of the sample fructose solution (any concentration) measure the O.D and read out the amount of glucose it contains from the standard graph.

Estimation of reducing sugars in fruits by Benedict's quantitative method

Benedict's method can be used to estimate the amounts of glucose, fructose, lactose and maltose. Certain amounts of Benedict's reagent is equivalent to certain amounts of the monosaccharides. This helps in their estimation. The Benedict's reagent contains $CuSO_4$, Na_2CO_3 and sodium citrate. The quantitative type of Benedict's reagent is used for this particular estimation.

Method

Take 25 gm of any ripe fruit, add a little distilled water and heat for 10 min. Grind the pulp well and filter through muslin. Add a pinch of lead acetate to precipitate the proteins, shake well and centrifuge for 10 min at 200 rpm. Take the supernatant and add a pinch of sodium oxalate to remove excess lead acetate and prevent further reduction. Centrifuge as before and remove the supernatant. Bring up the solution to 100 ml by adding distilled water.

Transfer the extract to a burette. Titrate 2.5 ml of Benedict's reagent (which is blue in colour) in a hot condition against the extract in the burette till the Benedict's reagent turns colourless. Repeat the titration to get a constant value.

Calculation

Calculate the quantity of different reducing sugars from the standard estimate of Benedict's reagent: 25 ml of the reagent is equivalent to 50 mg of glucose, 53 mg of fructose, 68.8 mg of lactose and 74 mg of maltose. If 2.5 ml of the reagent is equivalent to 0.005 gm glucose, then this is the quantity present in the titer value amount (i.e. the amount of the extract used up contains 0.005 gm of glucose).

Estimation of starch by modified method of Mc Cready *et al*

The starch present in cereal grains is solubilized with perchloric acid and treated with anthrone to determine the glucose content which is in turn used to determine the starch content. Prepare the anthrone reagent according to the procedure for estimation of glucose experiment given earlier.

Standard graph

Prepare and use the standard graph explained in the experiment of estimation of glucose by anthrone method.

Sample

Grind 1 gm of seed material in 10 ml ethanol. Heat in a water bath for 5 min, cool and centrifuge at 3000 rpm for 10 min. Take the precipitate and solubilize with 5 ml of 52% perchloric acid for 30 min and centrifuge at 3000 rpm for 10 min. Take the supernatant and make up to 10 ml with 52% perchloric acid. Take 0.5 ml and 1 ml of the extract, bring each to 5 ml with distilled water, add 5 ml of anthrone reagent and heat them in a water bath for 10 min till the green colour develops. Cool and measure optical density at 630 nm and reduce the glucose content from the standard graph. Convert the amount into starch content by content by multiplying with the glucose equivalent 0.9.

Estimation of soluble amino acids by ninhydrin method (phenyl alanine)

All α - amino acids react with ninhydrin, a powerful oxidizing agent, yielding a pinkish brown or purple colour due to the formation of a complex which can be measured by a spectrophotometer. Prepare the ninhydrin reagent by dissolving 0.5 gm ninhydrin in 100 ml of acetone.

Standard graph

Prepare the standard solution with 50 mg of phenylalanine in 100 ml of 1:1 ethanol and distilled water (500 µg /ml). Pipette out different volumes of the standard solution into a series of test tubes (containing 100 - 500 µg) and make them up to 5 ml with distilled water.

Add 2 ml of ninhydrin reagent and heat in a water bath for 15 min, until the purple colour develops. Cool the tubes to room temperature and measure the optical densities at 575 nm. Prepare the standard graph.

Sample

Proceed with the unknown sample solution (0.5 and 1 ml) in the same way as above and measure the optical density. Deduce the amount of the amino acid from the standard graph.

Estimation of proline in leaf material

Amino acids react with ninhydrin to give a blue or purple colour, but under extreme acidic conditions, proline may give a pinkish yellow colour which may be measured spectrophotometrically. Prepare ninhydrin reagent by dissolving 1.25 gm ninhydrin in a mixture of 2 ml of 0.6 N phosphoric acid and 30 ml of glacial acetic acid by heating gently.

Standard graph

Dissolve 10 mg of proline in 5 ml of isopropanol and bring it up to 500 ml with distilled water (20 µg / ml) to prepare the standard solution. Pipette out different volumes of proline solutions in a series of test tubes (2 - 20 µg) and bring each one up to 2 ml with distilled water. Add 2 ml of ninhydrin reagent and 2 ml of glacial acetic acid and heat the contents at 100°C for 1 hour in a water bath (carry this out in the fume hood). Stop the reaction by transferring the tubes to a cold water bath. Extract the mixture with 4 ml of toluene in each test tube by shaking vigorously for 20 seconds. Separate the toluene layer and warm it to room temperature. Record the optical density at 520 nm using toluene as blank. Prepare the standard graph.

Sample

Grind 0.5 gm of senescent leaves with 10 ml of 3% sulpho salicylic acid (w/v in distilled water) and filter through a Whatman filter paper. Take 2 ml of the extract containing proline and proceed as for the standard solutions. Measure the optical density and deduce the proline content from the standard graph and proceed to calculate its amount on fresh weight basis.

Separation of amino acids and identification by paper chromatography

The separation of amino acids over a stationary phase (paper) occurs due to the different rates of their movement. The solutes (amino acids) move along the chromatographic paper depending on their solubilities. The higher the solubility, the faster will be the movement. Hence hydrophilic amino acids migrate faster (farther) than do hydrophobic amino acids.

Solvent system

Prepare the solvent with n-butanol, acetic acid and distilled water in the ratio of 4:1:1. Shake the mixture well.

Reagent spray

Prepare the spray with 200 mg ninhydrin in 100 ml of acetone.

Standard amino acids

Prepare solutions of methonine, alanine, aspartic acid, proline and leucine by dissolving in isopropanol (10 mg in 10 ml of isopropanol).

Procedure

Take an appropriate sized chromatographic paper. Leave a margin of 2 cm from the bottom and spot each amino acid solution with a capillary tube repeatedly, drying the spot well after each application. Leave enough space between each spot on the base line and also ample margins on the sides of the paper. Roll up the paper with spots on the out side and pin it up. Pour the solvent in a beaker to a level below the base line and stand the paper cylinder in it. Cover with a bell jar or chamber and make it airtight. Leave it for 3 hours and then disassemble. Mark the solvent front with a pencil. Spray with ninhydrin after air drying the paper. The spots will appear (Fig-15).

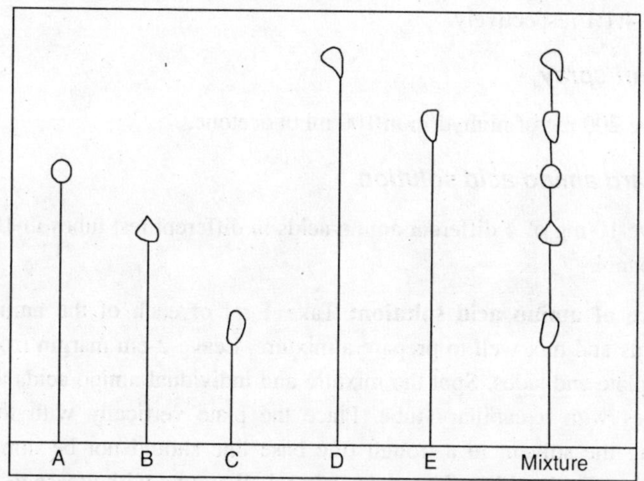

Fig 15: Separation of amino acids (A, B, C, D) by chromatography.

Measure the distance of the solvent run and the individual spots. Calculate the RF values.

$$RF = \frac{\text{Distance traveled by solute}}{\text{Distance traveled by solvent}}$$

Separation of amino acids by Thin Layer Chromatography (TLC)

The amino acids are separated on stationary phase of silica gel by moving along with the solvent, depending on their solubilities. Again the higher the solubility, the faster the movement.

Silica gel plate

Prepare a slurry by suspending 25 gm of silica gel in 50 ml of distilled water and by quickly shaking the suspension. Spread the slurry with a spreader (a broad spatula or a bent glass rod) immediately on a row of glass plates placed in a row. Let them dry and store them. Heat the TLC plates in an oven at 110°C to activate them before spotting with the amino acid solutions (Glass plates of 20 × 20 cm and 0.4 cm thickness will be appropriate).

Solvent systems:

Prepare the solvent system by mixing n-butanol, acetic acid and water in the ratio of 4:1:1 respectively.

Reagent spray

Dissolve 200 mg of ninhydrin in 100 ml of acetone.

Standard amino acid solution

Dissolve 10 mg of 4 different amino acids in different test tubes in 10 ml of isopropanol.

Mixture of amino acid solution: Take 1 ml of each of the amino acid standards and mix well to prepare a mixture. Leave 2-cm margin from base of the plate and sides. Spot the mixture and individual amino acids at equal distances with a capillary tube. Place the plate vertically with the base touching the solvent in a trough (the base line should not be immersed). Cover the trough with a glass plate and seal all around with grease to make it airtight. Let it stay for 3 hours and remove the plate. Mark the solvent front with a pencil very lightly. Spray with ninhydrin after drying the plate thoroughly.

Place the plate in an oven at 110°C for 5 min until completely dry. The spots appear (Fig - 15) on the plate. Calculate the RF value as explained earlier for each spot (amino acid) after tracing them out on tracing paper. Calculate the RF values in the mixture and compare them with those of individual values.

Estimation of total nitrogen (and proteins) by Micro-Kjeldahl's method

This method of estimation of nitrogen content is generally used for analysis of food material and agricultural samples. This also gives an estimate of both the nitrogen present in the proteins and the non-protein nitrogen of a sample. The nitrogenous compounds are converted to ammonium sulphate by digestion with H_2SO_4 in the presence of a catalyst. Ammonia liberated by steam distillation in the presence of a strong alkali is estimated. By finding out the amount of ammonia formed from a known amount of sample, the protein content can be calculated as follows: 1 mg nitrogen equals 6.25 mg protein.

Reagents

- 2% boric acid solution to which 20 ml of 0.1% bromocresol green and 4 ml of 0.1% methyl red are added.
- 0.01 N HCl solution.
- 40% (w/v) NaOH solution.
- Concentrated sulphuric acid.
- Catalyst (either a 1:2 mixture of $CuSO_4$ & K_2SO_4 or selenium dioxide).
- Standard 1% ammonium sulphate solution.

Apparatus

The Micro-Kjeldahl apparatus consists of a boiling water flask that produces steam which is bubbled into the sample solution mixed with the strong alkali (40% NaOH). Ammonia is produced which converts the boric acid to ammonium borate. This on titration with 0.01N HCl will yield the amount of ammonia.

Standardization

Conduct the experiment with known amounts of ammonium sulphate solution. Add 10 ml of ammonium sulphate solution and 10 ml of 40% NaOH into a Kjeldahl flask and start heating. The ammonia liberated gets fixed as ammonium borate when directed into the boric acid solution. Titrate this against 0.01 N HCl until the bluish green colour changes to pink and note the volume of acid used to calculate the amount of ammonia from the distillate. 1 ml of 0.01 N HCl is equivalent to 140 μg of nitrogen as present in ammonia. The amount of nitrogen is multiplied by 6.25 to arrive at the approximate protein content.

Sample

Place 20 mg dried seed powder in a Micro-Kjeldahl flask, add about 20 mg of the catalyst. Heat in a digestion rack for 3-4 hours or more till the solution becomes colourless. Add a few drops of H_2O_2 for clearing the solution. Make up to a known volume. Process this extract in the same way as the known ammonium sulphate standard and calculate the amount of total nitrogen as well as the protein content in the sample.

Estimation of proteins by Biuret method

Proteins containing two or more peptide bonds form a purple complex with copper salts in alkaline media, which can be measured spectrophotometrically.

Standard graph

Prepare a standard protein solution (1 gm/100 ml) and pipette out different volumes of the solutions (containing 1 to 10 mg) into a series of test tubes and add distilled water to make up to 1 ml. Add 5 ml of the Biuret reagent, shake well and let it remain for 30 min until the colour develops. Measure the optical density of the contents of all test tubes at 540 nm, using a blank without protein and plot the standard graph.

Unknown sample

Proceed with the sample solution (0.3 and 0.6 ml) in the same way as the standard solutions and measure the optical density. Deduce the protein content of the unknown sample by reading from the standard graph.

Estimation of protein content by Lowry's method (Lowry *et al.* 1951)

The final blue coloured complex formed by the Folin-Ciocalteau reagent is a result of (1) Biuret reaction of protein with copper ions in alkali. (2) Reduction of the phospho-molybdic and phospho-tungstic reagent by the tyrosine and tryptophan present in proteins. The colour is measured spectrophotometrically.

Reagents

- 2% Na_2CO_3 in 0.1N NaOH
- 0.5% $C_4SO_4.5H_2O$ in 1% Na.K Tartarate
- Alkaline copper solution: Mix 50 ml of reagent (1) with 1 ml of reagent (2) (prepare fresh)
- Folin-Ciocalteau (1N): Add 1.3 ml distilled water to 1 ml of the reagent.

Standard graph

Prepare a standard protein solution 5.6 mg bovine serum albumin dissolved in 25 ml of distilled water (200 µg/ml) and pipette out different volumes of the solution into a series of test tubes (containing 20 to 200 μg) and add distilled water to bring up to 1 ml. Add 4 ml of reagent (3) and mix well. Allow it to stand at room temperature for 10 min. Add 0.4 ml of Folin-Ciocalteau reagent and shake well immediately. Measure the optical density at 660 nm after the blue colour develops and prepare the standard graph.

Unknown sample

Grind 200 mg seeds with 10 ml distilled water. Keep the slurry for 1/2 hour for the soluble proteins to dissolve in the water. Centrifuge it for 10 min at 3000 rpm. Remove the supernatant and keep aside. Grind the precipitate again with a little distilled water, centrifuge and pool the supernatants. Add 5-10 ml of 20% trichloro acetic acid stirring constantly until a precipitate is formed. Centrifuge at 3000 rpm for 5 min. The precipitate contains the proteins. Dissolve it in 5 ml of 0.1 N NaOH and bring it up to 50 ml with distilled water (protein extract). Take 0.2 and 0.4 ml of the protein extract in different test tubes and proceed in the same way as for the standard solution. Measure the optical densities of the solutions after the development of blue colour (at 660 nm). Deduce the protein content from the standard graph. Calculate the protein content and present it as

$$\frac{\text{X - grams soluble protein}}{\text{per gram seeds.}}$$

Estimation of proteins by Bradford's method (Bradford, 1976)

This method is the most recent and accepted one. In this, the colour of coomassie brilliant blue G250 in dilute acid solution changes proportionally as the dye binds to protein. The reagent of Coomassie blue dye dissolved in phosphoric acid and methanol is available in a kit as Bio - Rad Protein Assay reagent purchased from Bio - Rad Laboratories, Richmond, CA, U.S.A.

The dye solution can also be prepared as follows: Mix 10 mg of Coomassie brilliant blue G-250 with 10 ml of 88% phosphoric acid and 4.7 ml of absolute ethanol, dilute the mixture to 100 ml with dH2O. The absorption at 550 nm should be around 1.18. Adjust with dye / water.

Standard graph

Prepare a standard protein solution with bovine serum albumin (BSA) (25 μg/ml). Pipette out different volumes of this solution into a series of test tubes (2.5, 5, 10, 15 and 20 μg) and add distilled water to bring to 1 ml. Add 0.2 ml of the dye reagent into each of the tubes and also a blank tube. Incubate for 5 min at room temperature and measure the colour (optical density) at 595 nm in a spectrophotometer using the blank. Plot a standard graph for O.D values and their respective protein concentration.

Unknown sample

Prepare the protein extract from the plant material as explained in the previous experiment and take 0.2, 0.4 and 0.6 ml of it in different tubes and proceed as for the standard sample. Note the O.D values at 595 nm and calculate the protein content in the sample from the standard graph. Express the value in mg protein per gram plant sample.

Determination of iso electric point of proteins (IP)

If an electric current is passed through protein solution, no migration occurs if that happens to be the iso electric point of the protein (IP). At pHs higher than I.P, protein micelles migrate towards the anode (negative pole) and at values below the I.P, they migrate towards the cathode (positive pole).

Sample

Any commercially available protein can be used. Use distilled water as the negative control.

Procedure

Prepare 9 beakers with 2 ml 0.1 N sodium citrate and 10 ml distilled water in each of them. Then add 0.1 N citric acid to them to establish different pHs in the different vessels (3.0, 3.5, 4.0, 4.5, 5.5 and 6.0 pH). Adjust the final volume of each solution to 18 ml with distilled water. Pour them into test tubes and label them. Add 1 ml of the protein solution to each of them slowly and mix well. Incubate them for 20 min at room temperature. Visually, check the turbidity in each test tube. The pH of the solution with maximum relative turbidity represents the iso electric point of the protein. If more than one solution exhibits the same turbidity, arrive at an approximate value. The solubility of protein varies with the pH.

Electrophoresis of proteins: SDS - PAGE of proteins [modified from Hames and Rickwood, 1987]

SDS - polycrylamide gel electrophoresis involves casting of gels, loading of protein samples, running the gels with electric power packs and staining of gels. Acrylamide is a neurotoxin. Handle it only with gloves and do not mouth pipette. Generally 10 - 15% gels are preferred for proteins. The electrophoresis gel constitutes a running gel with a stacking gel cast over it. The running buffer is prepared and the protein samples are loaded in wells before switching on the apparatus.

Requirements

- ***30% acrylamide (toxic ! use gloves !)***

  30 g acrylamide

  0.8 g bis-acrylamide

  Dissolve in distilled water and bring up to 100ml. Store at 4°C in brown bottle.

- ***Resolving buffer (3M) (pH 8.8) (Running gel buffer)***

  363 g Tris - base

  Bring up to 1 litre

- ***Stacking gel buffer (0.5 M)***

  59.8 g Tris - base

  Bring up to 1 litre

  pH 6.8

- ***Ammonium persulphate (APS) (10% solution, toxic !)***

  0.2 g ammonium persulphate

  Dissolve in 2 ml distilled water

  Store at 4°C

- ***Running (electrophoresis) buffer (8 X)(pH 8.6)***

  24 g Tris - base

  115 g glycine

  8 g SDS (sodium dodecyl sulphate)

  Bring up to 1 liter with distilled water

- ***SDS gel - loading buffer (1 X)***

50 mM Tris - Cl (pH 6.8)

100 mM dithiothreitol (DTT)

2% SDS

0.1% bromophenol blue

10% glycerol

(1 X SDS gel - loading buffer lacking dithiothreitol can be stored at room temperature. Dithiothreitol should then be added, just before the buffer is used, from a 1 M stock).

- ***Protein extraction buffer (pH 8.0)***

  5% sucrose

  4% SDS

  1% 2 - mercaptoethanol

  0.05 mM EDTA

  20 μM leupeptin

- 1mM phenyl methyl sulfonyl fluoride (PMSF)
- 0% sodium dodecyl sulphate (SDS)
- N, N, N, N - tetra methyl ethylene diamine (TEMED)
- 0.5 M NaOH
- 0.5M HCl

Protein extraction from sample

Grind the sample (leaf material) with the extraction buffer and precipitate the protein with 4 volumes of acetone (1hr at -20°C) and resuspend in 0.5M NaoH (2 hr at 37°C). Neutralize with 0.5M HCl and measure the concentration of protein. 5 g of protein samples can be used for the SDS-PAGE.

Casting of the gel

- ***Running gel (10%)***

  2 ml 30% acrylamide (with 0.8% bis-acrylamide)

  0.75 ml resolving buffer

  3.2 ml distilled water

  0.06 ml 10% SDS

0.06 ml 10% APS

0.006 ml TEMED (add just before loading)

- **Stacking gel**

400 μl 30% acrylamide (with 0.8% bis-acrylamide)

500 μl stacking gel buffer

1.06 ml distilled water

40 μl 10% SDS

0.02 ml 10% APS

0.002 ml TEMED (add just before loading)

Procedure

1. Assemble the gel plates with spacers according to the manufacturer's instructions. Determine the volume of gel mold and prepare the equivalent amount by scaling up the amounts of the ingredients given above.

2. Quickly add together the ingredients of the running gel in a flask (do not allow air bubbles to form). Pour this quickly into the cavity in between sealed glass plates. Pour water saturated iso propanol over it to get an even surface. However, before pouring the gel, seal the bottom spacer by using a small amount (250 μl) of running gel with 0.006 ml TEMED (the high concentration of TEMED will solidify it instantly). Leave sufficient space for the stacking gel.

3. Layer a thin film of 95% -100% ethanol for an even surface. After the running gel solidifies remove the ethanol by pouring it out and using a piece of filter paper.

4. Mix together the ingredients of the stacking gel and pour it over the solidified running gel. Place a comb for the formation of wells. Maintain 1 cm distance distance between the bottom of comb (wells) and the surface of running gel.

Preparation of samples and loading

1. Denature the proteins by heating the samples to 100° C for 3 min in the gel-loading buffer.

2. Mount the gel in the electrophoresis apparatus and add the electrophoresis buffer to the top and bottom reservoirs Remove air bubbles if any. Remove the comb and wait till the wells fill with the buffer before loading 15 µl of each sample in a predetermined order into the bottom of the wells with a Hamilton syringe. Load the protein markers in the first well.

Running the electrophoresis unit

1. Attach the apparatus to the power supply (connect the positive electrode to the bottom of the buffer reservoir. Apply a voltage of 10 - 15 V/ cm to the gel.

2. Run until the bromophenol blue front reaches the bottom of the gel. Place the gel plates on paper towels and gently pry them open with a spatula. The gel must now be fixed and stained before viewing and photographing.

Staining the gel

1. Stain it for 2 - 3 hrs in coomassie blue stain mixture containing 50 ml methanol, 10 ml acetic acid 40 ml distilled water and 0.1% coomassie blue stain. The gel can be placed in a container kept on a slowly rotating platform.

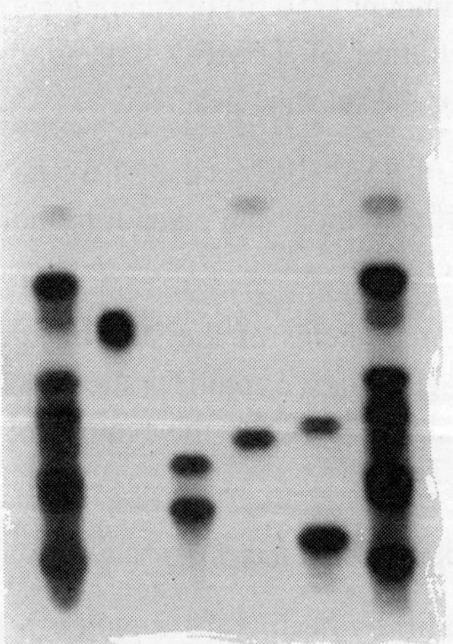

Fig: 16. Polyacrylamide gel electrophoresis.

2. Destain the gel in the destaining mixture (same as above without the stain) for about 5-6 hours and photograph the bands of separated proteins (Fig-16). Record the RF values and molecular weights of the proteins.
3. The gel can be stored in water containing 20% glycerol at 4°C. The gel can also be used for blotting (Western blotting).

Determination of iodine number of oils

The iodine number reflects the degree of unsaturation of a fat or oil. It is the amount of halogen (I) expressed as grams of Iodine absorbed by 100 gm of fat or oil. Unsaturated fatty acids will react with halogens by addition at the double bond.

Requirements

- Any oil
- ***Hanus Iodine solution:*** Dissolve 15.2 gm of Iodine in 1000 ml of glacial acetic acid. Add 2 - 3 drops of bromine to it. Prepare 15% potassium iodide (KI) solution (w/v in water). Prepare 0.1 N sodium thiosulphate and 1% starch solution.

Procedure

Take 1.2 ml of any oil in an Erlenmeyer flask and add 10 ml of chloroform (**careful! carcinogen**!) and 30 ml of Hanus Iodine solution (perform in hood). Keep the flask in complete darkness (by wrapping in foil) for 1/2 hour with frequent rotations. Add 10 ml of 15% KI solution and 100 ml of boiled and cooled water and titrate this mixture against 0.1 N sodium thiosulphate from a burette. Note the point when the solution turns pale yellow. Add 2 ml of 1% starch. The solution will turn blue. Continue titration and record the end point when the blue colour disappears. Titrate in the same way with a blank and record the end point value. Note the difference in the titre value (sample and blank). Calculate the iodine number with the help of this difference by relating it to the equivalent of 1 ml of sodium thiosulphate to 6.325 gm of Iodine.

Determination of saponification number of oil

Alkaline hydrolysis of fats is called saponification since one of the products of the reaction is soap. It is defined, as the number of mg of KOH needed to saponify 1 gm of fat. Since each molecule of fat regardless of its size requires 3 molecules of KOH to saponify it, the saponification number really indicates the number of fat molecules per gram of fat. The larger the molecule, the fewer will be the molecules per gram of fat. The saponification number therefore becomes a measure of the average molecular weight of the fatty acid in a triglyceride and of the fat itself. Saponification number is an inverse mean of molecular weight of the mixed glycerides composing a given fat or oil.

Requirements

- Any oil.
- 0.5 NH_4Cl
- 0.5 N alcoholic KOH

Procedure

Weight out 3 - 4 gm of any oil in each of two clean dry 100 ml round bottomed flasks (replicates). Add 50 ml of 0.5 N alcoholic KOH to each flask and also into a third flask, which is used as a blank. Place 2 - 3 glass beads in the three flasks and connect the flasks to reflex condensors and heat on a hot plate for 30 minutes. Remove the reflex condensor cool the flask and add 8 drops of phenopthalein indicator to each flask and titrate with 0.5 N HCl in the burette. Record the titre values. The difference in titre values of sample and blank represents the excess of KOH that was not used to saponify the oil. Convert this to milli-equivalents of KOH by multiplying with 0.5. Convert to gram equivalent of KOH. 1 gm = 56.1 gm (equivalent weight of KOH). Find the amount of KOH required to saponify 1 gm of oil, which is the saponification number of the oil.

Estimation of phosphate

Phosphorus in leaf material is converted to phosphates on addition of an acid mixture. When ammonium molybdate - ammonium vanadate solution is subsequently added, a yellow colour results. The intensity of the colour is generally measured spectrophotometrically.

Requirements

- *Acid mixture:* Mix 750 ml of concentrated HNO_3, 150 ml concentrated H_2SO_4 and 300 ml of 60% perchloric acid.

- ***Ammonium molybdate - vanadate solution:*** Dissolve 25 gm of ammonium molybdate in 500 ml of distilled water and 1.25 gm of ammonium vanadate in 50 ml of 0.1 N HNO_3. Mix both just before use.
- 2N HNO_3

Standard graph

Dissolve 110 mg KH_2 PO_4 in 1000 ml distilled water. This contains 25-ppm phosphorus (25 μg/ml or 25-mg/100 ml). Pipette out different volumes (containing 25 to 150 μg) of the solution in a series of test tubes and bring them up to 8 ml with distilled water. Add 2 ml of 2N HNO_3 to each test tube followed by 1 ml of ammonium molybdate - vanadate solution. The yellow colour develops after 20 minutes. Measure the optical density at 420 nm using a blank and prepare a standard graph.

Sample

Take 1 gm of leaf material in a digestion flask. Add 10 ml of acid mixture to it in 4-5 instalments and heat the contents for 2 - 3 hours at 50° C. Transfer the material to a clean flask and make it up to the mark with distilled water. Filter the phospate extract and proceed in the same way as described for standard solution by pipetting out 0.5 and 1 ml. Deduce the phosphate content from the standard graph, after measuring the O.D.

Estimation of RNA

Pentose sugars react with HCl to produce furfural. Furfural reacts with orcinol in the presence of a catalyst ($FeCl_3$) to produce a green coloured complex that can be measured spectrophotometrically.

Requirements

- ***Orcinol reagent:*** Dissolve 1 gm of orcinol in 100 ml of concentrated HCl. Add 0.5 gm of Fe Cl_3 prior to the addition of orcinol.
- 0.5 N perchloric acid

Sample

Prepare standard RNA solution by dissolving 100mg yeast RNA in 100 ml of 0.5 N perchloric acid. Pipette out different volumes (containing 100 - 1000 μg) of the RNA solution in a series of test tubes and make up the volume to 4 ml with distilled water. Add 4 ml orcinol and heat in a water bath for 20 min till the green colour develops. Record the optical densities at 660 nm (after cooling) with a blank and prepare the standard graph.

Sample

Grind 1 gm of leaf material with small amounts of ethanol in a mortar with pestle. Filter the homogenate through cheesecloth and centrifuge at 3000 rpm for 10 min. Collect the clear supernatant and make it up to 10 ml with ethanol. Remove the chlorophyll pigments by shaking the extract with 10 ml chloroform and collecting the colourless layer. Add 5 ml of 0.5 N perchloric acid to the extract and heat it for 10 min in a boiling water bath. Pipette out 0.5 and 1 ml of this extract and proceed in the same way as for the standard solutions. Record the O.D and deduce the RNA content from the standard graph.

Estimation of DNA by diphenyl amine method

The deoxyribose in DNA in the presence of acid forms hydroxy levulinic aldehyde, which reacts with diphenylamine to give a blue colour.

Requirements

- ***Diphenylamine reagent:*** Dissolve 1 gm of diphenylamine in 100 ml glacial acetic acid and by subsequently adding 2.5 ml concentrated H_2SO_4. This reagent should be prepared fresh before use. Just prior to use, add 0.25 ml of 1.6% acetaldehyde (or 0.5 ml aqueous acetaldehyde).
- 0.1 M sodium phosphate buffer (pH 7.0)
- 1 N perchloric acid
- Calf thymus DNA
- Ethanol
- Young, freshly harvested leaves

Standard graph

Dissolve 40 mg of calf thymus DNA in 100 ml of 0.1-M sodium phosphate buffer. Stir at 4°C until dissolved and refrigerate it. Prepare working standards (of 0.2 mg/ml) from this by mixing equal volumes of the stock and 1N perchloric acid and heating for 15 min at 70°C before use. Pipette out different volumes of the standard solution (containing 20 - 200 μg) in a series of test tubes and make up to 3 ml with distilled water. Add 2 ml of the diphenylamine reagent and heat for 10 - 15 minutes at 90°C until a blue colour develops. Cool and measure the optical densities at 600 nm against a blank. Draw the standard graph.

Sample

Grind 1 GM leaf material in 5 ml of ethanol. Filter the homogenate through cheesecloth and make it up to 10 ml with ethanol. Add equal quantity of 1N perchloric acid and heat for 10 min at 90°C. Centrifuge at 1000 rpm for 10 min. Take the supernatant and make up to 10 ml with 1N perchloric acid. Pipette 0.5 and 1 ml of the DNA extract into test tubes and proceed as for the standard solutions. Deduce the DNA content (after measuring the O.D) from the standard graph.

Estimation of ascorbic acid in fruit

The dye 2, 6 dichlorophenol indophenol is blue in alkaline conditions and pink in acidic conditions. The blue colour changes to pink by reduction with ascorbic acid.

Requirements

- ***Metaphosphoric acetic acid mixture:*** Add 15 gm of metaphosphoric acid to 490 ml of dilute acetic acid (45 ml acetic acid plus 450 ml distilled water). Prepare the dye by dissolving 34 mg of sodium bicarbonate and 104 mg of 2,6 dichlorophenol indophenol in distilled water and make it up to 200 ml with distilled water.
- ***Standard ascorbic acid solution:*** Dissolve 10 mg of ascorbic acid in distilled water and make it up to 100 ml in a standard flask.
- 2, 6, dichlorophenol indophenol dye

Sample

Weigh a lemon fruit. Then squeezes out the juice from the lemon fruit and make it up to 45 ml with distilled water and add 5 ml of metaphosphoric acid.

Procedure

The standard solution and sample solution are estimated in an identical manner. Take 1 ml of the solution in an Erlenmeyer flask and add 9 ml of metaphosphoric acid. Titrate this against the 2,6-dichlorophenol indophenol dye. Titre values are noted when the blue colour turns to pink. The amount of ascorbic acid in the sample can be calculated by taking into consideration the amount of dye consumed by the known standard solution. Present the value as a percentage.

Determination of total and titrable acidity

Plant cells are acidic. Acidity is a measure of H^+ ions, which dissociate easily. Weak acids have a low concentration of H^+ ions and dissociate slowly. But both require equal amounts of alkali for neutralization. This is total acidity. Titrable acidity includes the total acidity and also the potential H+ ions (i.e. the concentration of H+ ions already present but demonized by OH^- ions). The true acidity is expressed as the gm moles of H^+ ions and the titrable acidity as normality of acid.

Requirements

- 0.1 N NaOH
- 0.1 N HCl
- Bromomethyl indicator
- Phenopthalein indicator
- Fresh young leaves (e.g. Oxalis)

Procedure

Grind 5 gm Oxalis leaves in a little distilled water. Filter through cheesecloth and centrifuge at 100 rpm for 15 min. Take the supernatant and make up to 100 ml. Titrate 10 ml of the extract with 2 - 3 drops of bromo methyl indicator against 0.1N NaOH. Note the end point when the pale yellow colour changes to light green. Now add 2 drops of phenopthalein indicator and titrate against 0.1 N HCl. Not the end point when the pale yellow reappears.

Calculate the normality of the extract from the normality and volume of NaOH according of the formula N1 V_1 = N2 V_2. Calculate the titrable acidity in terms of normality of the extract in the second titration from the normality and volume of the HCl according to the formula N1 V_1 = N2 V_2. Calculate total acidity by adding normality of the extract (against NaOH) to the normality against HCl.

Estimation of total chlorophyll content and ratio of chlorophyll *a* and *b*

Chlorophyll *a* and *b* have different absorption peaks. The spectrophotmetric readings of the absorption peaks give an estimate of the concentration of these pigments in the leaves.

Requirements

- Fresh young leaves
- Acetone

Procedure

Grind 1 gm of green leaf material and homogenize it in 10 ml acetone. Centrifuge the homogenate at 3000 rpm for 10 min. Transfer the supernatant to a 100 ml standard flask. Repeat the tissue extraction with acetone until the extract is free from pigments. Make up to 100 ml with acetone. Record the optical density in a spectrophotometer at 645 and 663 nm.

Calculate chlorophyll a, b and total chlorophyll contents with the following formulae:

Total chlorophyll: O.D (645) × 20.2 + O.D (663) × 8.02

Chlorophyll a: O.D (663) × 12.7 — OD (645) × 2.69

Chlorophyll b: O.D (645) × 22.9 — O.D (663) × 4.68

The total amount of the pigments present in the leaf material can be calculated as follows

$$\text{Total amount of pigment} = \frac{V}{W} \times 1000 \text{ mg/g fresh Wt.}$$

where V = volume of the total extract and W = fresh weight of leaf material.

Estimation of carotenoids

Carotenoids are C40 terpenoids and are widely distributed in plants. They function as accessory pigments in photosynthesis and as colouring pigments in flowers and fruits. The extraction should be carried out in diffuse light to avoid their breakdown.

Requirements

- 60% aqueous KOH
- Petroleum ether
- Ethanol
- Methanol / acetone
- Any dark green leafy vegetable (e.g. spinach)

Procedure

Grind 2 gm of the leaves in 20 ml of acetone or methanol. Decant the extract into a flask and re-extract three times with 20 ml of the acetone / methanol. Add 40 ml petroleum ether to the extract in a separatory funnel. Add water if needed to separate the two phases. The carotenoids will be present in the top ether phase. Collect and evaporate ether by heating in a hot water bath at 35°C in an exhaust hood. Dissolve the residue in a small amount of ethanol. Add 60% aqueous KOH at the rate of 1 ml for every 10 ml of the ethanol extract. This will remove the chlorophylls and lipids and cleave esterified carotenoids. Leave it overnight and add equal amount of water. Partition twice with petroleum ether, collecting the ether phase. Evaporate and dissolve the residue in a minimum quantity of ethanol. Measure the absorbance at 450 nm. Compare with the calibration curve of β-carotene for estimation or follow the procedure below. Measure the absorbance at 450 and 670 nm.

Calculate the amount of carotenoids present in the material by the following method:

$$\text{C (carotenoids in mg)} = \frac{\text{D X V X F X10}}{2500}$$

where D = absorbance at 450 nm; V = volume of extract; F = dilution factor (if diluted by making up the solution to a known volume); 2500 = average extinction coefficient of the pigments.

Chemical separation of chloroplast pigments

Chloroplast pigments can be separated by their differences in solubilities.

Requirements

- 80% acetone

- 92% methanol
- 30% methyl alcoholic potash
- Petroleum ether
- Diethyl ether
- Fresh leaves (from any common plant like sunflower, rice or tecoma)

Procedure

Grind 10 gm of fresh leaves with a pinch of glass powder and by adding 110 ml of 80% acetone. Take 40% of the acetone extract in a separatory funnel. Add 50 ml of petroleum ether and shake gently. Add 10 ml of distilled water. Two liquid phases should separate out. Discard the lower layer. Add 50 ml of water to the upper green layer and shake gently. Two layers should separate out. Discard the lower layer. Repeat this twice. Add 50 ml of 92% methanol and shake gently. Two phases should separate out. The lower methanol layer contains chlorophyll *b* and xanthophylls. Set it aside. The upper petroleum ether layer contains chlorophyll 'a' and carotene.

Pour 30 ml of the upper layer in a 100 ml graduated cylinder and add 50 ml of fresh 30% methyl alcoholic potash solution. Shake well and set aside for 10 min. Add 30 ml of water and shake to separate two phases. The lower water alcohol layer will be bluish green and will contain chlorophyll *a*. The upper petroleum ether layer will be light yellow and will contain carotene (Fig-17a).

Pour 50 ml of the methanol layer (lower layer) into a separatory funnel and add 50 ml diethyl ether. Shake gently. Add 50 ml of distilled water to separate the lower water wash layer. Discard the layer and repeat the washing 5 times. Now two layers separate out: An upper diethyl ether layer with the pigments and a lower methanol layer. Discard the methanol layer and pour the upper layer in a 100 ml graduated cylinder. Add 15 ml of fresh 30% methyl alcoholic potash slowly. Shake gently. Add 30 ml of water which causes the phases to separate. The lower water layer will be grass green and will contain chlorophyll 'b' and the upper ether layer will contain xanthophyll and appear dark yellow in colour (Fig. 17b).

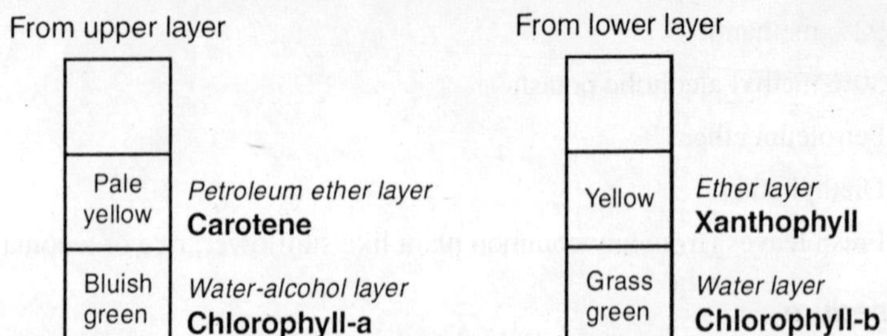

Fig: 17 (a & b). Chemical separation of chloroplast pigments.

Column chromatography of photosynthetic pigments

The distribution rates of different pigments over a column will differ due to their differences in solubility. The pigments separate out as bands as the extract moves down the column and can be collected separately to find their absorption maxima spectrophotometrically. The order of collection (emergence) from the column is carotenes, xanthophylls, chlorophyll *a* and chlorphyll *b* in sequence.

Requirements

- Cellulose for preparing the column
- Petroleum ether
- Acetone
- Leaf material

Procedure

Prepare a column by pushing a cotton plug to the bottom of a burette. Connect the nozzle to a vacuum pump. Prepare a slurry of cellulose powder mixed with petroleum ether and pour into the column. Tap gently to settle the cellulose without forming any gaps. Keep the surface of the column on top moist with petroleum ether. Prepare the pigment extract by grinding 10 gm of leaf material with acetone. Transfer 40 ml of extract after filtration through cheesecloth into a separatory funnel. Add 50 ml of petroleum ether, shake gently and add 10 ml of distilled water. Two phases will separate out. Draw off the lower acetone-water layer and discard it. Pour the upper petroleum ether layer into a beaker. About 5 ml of this pigment extract will serve as the sample. Load the sample over the column and start the vacuum pump. After the sample has moved into the column, pass a solvent containing 100 parts petroleum ether and 20 parts acetone through the column. Eventually, four bands of pigments will appear. The suction force gradually brings the pigments to the nozzle tip so that they can be collected separately. Measure the optical density of each pigment at different wavelengths using the solvent as blank. Calculate the absorption maxima for each one by plotting the absorption spectra.

Determination of water potential of tissue by falling drop method

Plant tissue loses water to solutions having a lower water potential thus diluting them. The tissue gains water from solutions with higher water potential thus concentrating them. Sucrose solutions are used for determining water potential. The changes can be measured by a refractometer. However, the dye method is a simple and inexpensive method and gives accurate results (Fig-18).

Requirements

- Graded sucrose solutions (0.15, 0.20, 0.25, 0.30, 0.35 and 0.40 molal)
- A single large potato
- Methylene blue

Procedure

Set up a series of test tubes with a graded sucrose solution (0.15, 0.20, 0.25, 0.30, 0.35 and 0.40 molal) in duplicate. Use one tube of each pair as control and the other for tissue incubation. Add one drop of methylene blue in each of the control tubes and mix well. Prepare potato plugs with a cork borer from a single potato and incubate them in the graded tubes (one in each concentration) for 15 minutes. Remove the tissue plugs from all the tubes, the contents of which should now be tested. Introduce one drop of solution into each tube from its respective control (the duplicate into which the methylene blue was added). The drop should be introduced into the test solution well below the surface.

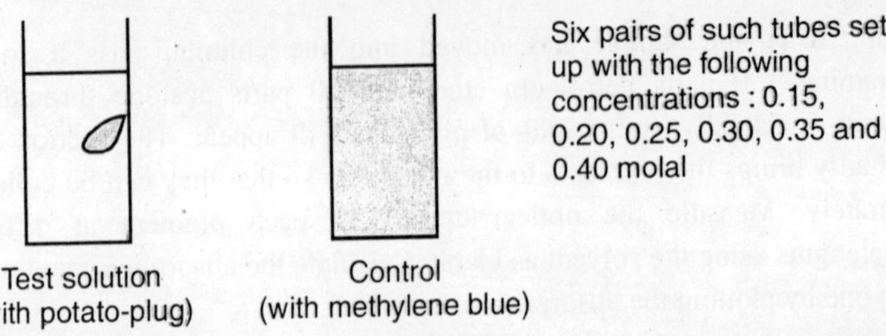

Fig 18: Determination of water potential by the falling drop method.

The three possibilities are:

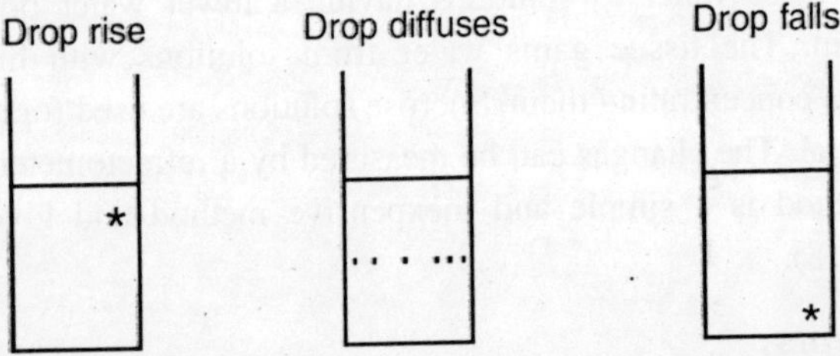

If the drop rises, it means that the test solution has become more dense after the incubation (due to the diffusion of water from the test solution into the tissue). If the drop falls, it means that the test solution has become lighter (due to the diffusion of water from the tissue into the solution).

Drop rises = The tissue had more negative water potential initially.

Drop falls = The test solution had more negative water potential initially.

Drop diffuses = Indicates no change. The water potentials were equal.

It can be concluded from the above experiment that the water potential of the potato tissue is the same as the test solution in which no change had occurred (drop diffused). The water potential of the potato can then be reported.

Estimation of indoleacetic acid (IAA)

Indoleacetic acid (IAA) is oxidized by ferric chloride in the presence of perchloric acid. IAA forms a complex with Salkovsky reagent, which appears pink in colour.

Requirements

- 35% perchloric acid

- ***Salkovsky reagent:*** Mix 1 ml of 0.5 M ferric chloride with 50 ml of 35% perchloric acid.
- Indole acetic acid (IAA)
- Ethanol

Standard graph

Prepare a standard solution (50 µg/ml) by dissolving 5 mg of IAA in 5 ml of ethanol and making it up to 100 ml with distilled water. Pipette out different volumes (containing 10 - 60 µg) of the standard solution into a series of test tubes and make each of them up to 4 ml with distilled water. Add 4 ml of Salkovsky reagent and incubate for 1/2 hour in the dark at room temperature. The solutions develop a pink colour. Measure their optical densities at 530 nm by using a blank. Prepare a standard graph.

Unknown sample

Pipette out 0.5 and 1 ml of the unknown solutions of IAA and proceed in the same way as for the standard solutions. Calculate the amount of IAA present in the unknown solutions by deducing the amount from the standard graph.

Determination of IAA oxidase activity

Plant tissues contain an enzyme, IAA oxidase, which inactivates excess of IAA produced in certain situations.

Requirements

- ***Tang and Bonner's reagent :*** Add 15 ml of 0.5 M $FeCl_3$ to 500 ml water and 300 ml concentrated H_2SO_4. Heat it to dissolve.
- ***2,4 - dichlorophenol 2 x 10^{-4} M*** : Dissolve 8 mg dichlorophenol in 250 ml distilled water.
- 0.1 M phosphate buffer (pH 6.2)
- Dry peas

Standard graph

Prepare 4 mg/100 ml IAA solution in distilled water. Pipette out different volumes of the solution in a series of test tubes (0 - 40 µg/ml) and make up to 2 ml with distilled water. Add 8 ml of Tang and Bonner's reagent and incubate for 30 minutes for the colour development. Measure the optical density at 420 nm with a blank and prepare the standard graph.

Sample enzyme extract

Soak peas overnight and place them on moist filter paper to germinate. Take 8-day-old seedlings and cut the seedlings into upper stem, lower stem and roots. Discard the cotyledons. Take 2 gm lots of each part and grind in 20 ml of 0.1 M KH_2PO_4 (pH 8.0) buffer in cold. Centrifuge for 10 minutes at 1000 rpm. Take the supernatant as the enzyme extract. Prepare 8 test tubes to contain 2 ml of 100 ppm IAA (200 μg/ml), 2 ml of 2,4 dichlorophenol (2 × 10-4 M) and 6 ml of phosphate buffer. Keep two tubes for each plant part and two tubes as blank. Add to the two tubes of each set 8 ml of the enzyme extract. Add 8 ml of buffer to the blanks. Incubate at 25° C. After suitable time intervals (0, 10, 20, 40 and 60 minutes) withdraw 2 ml of the contents of each tube and add 8 ml of Tang and Bonner's reagent and incubate for 30 minutes. Measure optical densities of all the samples with the blanks at a setting of 420 nm. Deduce the IAA content from the standard graph and interpret the results. The amount of IAA gives us an estimate of the amount of destruction carried out by the IAA oxidase during the incubation of each extract. Compare the values in the different plant parts.

Determination of amylase activity

Amylase is a hydrolytic enzyme which breaks down many polysaccharides (e.g. starch which is a polymer of glucose units linked together) to yield the disaccharide (maltose) as the end product.

$$\underset{\text{(Substrate)}}{\text{Starch}} + nH_2O \longrightarrow \underset{\text{(Product)}}{n\ \text{(Maltose)}}$$

Both substrate and product are colourless. The starch-iodine complex (blue coloured) cannot be used for a quantitative test. Hence the maltose can be estimated with dinitrosalcylic acid reagent.

Requirements

- Maltose
- Starch
- Barley seeds

- ***Dinitrosalicylic acid reagent:*** Dissolve 1 gm of 3,5-dinitrosalicylic acid, 30 gm of sodium potassium tartarate and 1.6 gm of sodium hydroxide in distilled water. Make it up to 100 ml with distilled water.

Standard graph

Prepare a 10 ml standard solution of maltose (1 mg/ml) and pipette different volumes of 2 ml with distilled water. Add 2 ml of the reagent to each tube. Heat the tubes in boiling water bath for 10 minutes. The solutions develop an orange-red colour. Cool the tubes and add 10 ml of water to each. Measure the optical densities at 520 nm and prepare the standard graph.

Enzyme extract

Germinated barley seeds (radicle protruded) can be used for the source of the enzyme amylase. Grind 10 gm of seeds in phosphate buffer (pH = 6) under cold conditions. Filter the homogenate and centrifuge for 5 minutes at 100 rpm (preferably in a refrigerated centrifuge) and keep the supernatant enzyme extract in an ice bucket.

Substrate: Prepare 1% starch solution and use it for a substrate.

Procedure

Pipette 2 ml and 2.5 ml of starch solution into different tubes. Add 2.5 ml of 0.1 N phosphate buffer (pH 6.7) to each. Add 1 ml of 1% NaCl to each and incubate the tubes for 10 minutes at 37° C. Now add 2 ml of enzyme extract and 1 ml of distilled water to each. Incubate the tubes for 15 minutes at 37° C and stop the reaction by adding 0.5 ml 2 N NaOH. Now add 0.5 ml dinitro salicylic acid reagent and mix well. Heat the tubes in a boiling water bath for 5 min. Cool the tubes and measure their optical densities. The amount of maltose formed in 15 min by the enzyme activity is equivalent to the optical density value. Compare with the standard graph and deduce the amount of maltose (mg) formed per ml of enzyme extract. However enzyme activities should always be expressed in terms of specific activity, which is defined as the amount of substrate utilized, or product formed per mg protein per minute at a definite temperature. Hence it will be appropriate if the protein concentrations of the barley seeds (enzyme source) is also determined.

Determination of catalase activity (Permanganate titration method)

Catalase is a haeme - containing protein with a Fe-porphyrin as its prosthetic group. Catalase is found in almost all living cells. It is a detoxifying agent for H_2O_2 and is extremely efficient in breaking H_2O_2 down to H_2O and O_2.

Requirements

- 0.1 M phosphate buffer (pH 6.8)
- 0.1 M H_2O_2
- 0.7 N H_2SO_4
- 0.01 N $KMnO_4$
- Fresh young leaves

Procedure

Grind 500 mg of leaves in 20 ml of the phosphate buffer in a pre-chilled mortar with a pestle. Filter the homogenate through cheesecloth and centrifuge it at 15,000 rpm for 15 min. Bring up the extract to 25 ml with the phosphate buffer. Place 1 ml of this extract in a test tube and add 2 ml of 0.1 M H_2O_2 and 3 ml of phosphate buffer. Stop the reaction by adding 1 ml of 0.7 N H_2SO_4 (or 10 ml of 2% H_2SO_4) after 5 min. Incubate for 5 min at 27° C and titrate the residual H_2O_2 against 0.01 N $KMnO_4$ (taken in the burette). Note the endpoint when the pink colour appears and persists for 30 sec. Repeat it till concurrent values are obtained.

Titrate a blank (take 1 ml of enzyme extract and 2 ml of H_2O_2 and add 1 ml of 0.7 N H_2SO_4 immediately and 3 ml of phosphate buffer) and record the titer value. The amount of H_2O_2 destroyed by catalase is calculated by the formula:

$$\frac{25 \text{ X } 0.85}{2} \text{ X } \frac{V}{W}$$

where, W = Weight of material used; V = Volume of $KMnO_4$ utilized (Blank sample value)

Present the catalase activity as enzyme units per gm leaf material. One unit of catalase is defined as that amount of enzyme, which breaks down 1 µ mol of H_2O_2/ min.

Determination of polyphenol oxidase activity

Polyphenol oxidase oxidizes phenol to quinone in the presence of pyrogallol, which is used as the substrate instead of polyphenols. Pyrogallol is converted to purpurogallin by polyphenol oxidase.

Requirements

- ***0.1 M pyrogallol:*** Dissolve 1.26 gm pyrogallol in 10 ml methanol and make it up to 100 ml with distilled water.
- ***0.02 M phosphate buffer (pH 7.0):*** Dissolve 2.722 gm of K_2HPO_4 in 100 ml of water to make 0.02 M K_2HPO_4. Then add 0.02 M KH_2PO_4 (3.48 gm KH_2PO_4 in 100 ml of distilled water) in the proportion of 206 ml and 94 ml respectively.
- ***2.5 M H_2SO_4:*** Dilute 70.5 ml concentrated H_2SO_4 to 1000 ml with distilled water.
- ***Enzyme extract:*** Grind 500 mg leaf material in 30 - 40 ml phsophate buffer (0.02 M) filter through cheese cloth and centrifuge at 2000 rpm for 10 min. Make up the extract to 100 ml with the buffer.

Procedure

Make up the reaction mix by adding 2 ml of buffer, 1 ml of pyrogallol and 1 ml of enzyme extract. Incubate for 5 min and stop the reaction by adding 1 ml of 2.5 N H_2SO_4. Record the optical density at 420 nm against a blank containing 1 ml of H_2SO_4, 2 ml of buffer, 1 ml of pyrogallol and 1 ml of boiled enzyme extract. Calculate the enzyme activity by subtracting the absorbence value of blank from the sample and express the enzyme activity as absorbing units per 1 gm fresh weight per 5 minutes.

Determination of peroxidase activity

Peroxidase catalyses the oxidation of various hydrogen donors like P-cresol, benzidine, guaiacol, ascorbic acid, nitrite and cytochrome C in the presence of H_2O_2. Peroxidases are present in all plant species. The enzyme has iron as prosthetic group, which can be dissociated readily. The tissue enzyme extract when added to pyrogallol in the presence of H_2O_2 is oxidized to a coloured derivative, which can be measured in a spectrophotometer.

Requirements

- ***Enzyme extract:*** Prepare the extract by using the method suggested for the polyphenol oxidase.
- 0.05 M pyrogallol in 0.1-M phosphate buffer (pH 6.0).
- 1% H_2O_2.

Procedure: Pipette 3 ml of pyrogallol phosphate buffer and 0.1 ml of enzyme extract into a cuvette. Add 0.5 ml of H_2O_2 and shake well. Measure the absorbence after 3 minutes at 420 nm. Record the value of the blank with boiled enzyme extract. Calculate the enzyme activity by subtracting the absorbence value of the blank from the sample and express the enzyme activity as absorbing units per 1 gm fresh weight per 3 min.

REFERENCES

1. Bradford. M. M. (1976) A rapid and sensitive method for the quantitation of microgram quantities of protein utilizing the principle of protein dye binding. Anal. Biochem. 72, 248-254.
2. Bruening. G., Giddle. R., Preiss. J. and Rudert. F. (1970) Biochemical experiments. John Wiley and Sons. New York.
3. Clark. J. M. Jr and Switzer. R. L. (1977) Experimental Biochemistry. 2nd edition. W. H. Freeman. San Francisco.
4. Hames. B. D. and Rickwood. D. (1987) Gel electrophoresis of proteins. A practical approach. IRL press. Oxford.
5. Lowry. O. H., Rosebrough. N. J., Farr. A. L. and Randall. R. J. (1951) Protein measurements with the Folin - phenol reagent. J. Biol. Chem. 193, 265 - 275.
6. Methods of Enzymology...Serial publications.
7. Plummer. D. T. (1977) An introduction to practical Biochemistry. 2nd edition. Tata - McGraw Hill Bombay.
8. Smith. I. and Feinberg. J. G. (1973) Paper and thin layer chromatography and electrophoresis. 2nd edition. Longmans. London.
9. Williams. B. L. and Wilson. K. (1975) Editors. Principles and techniques of practical Biochemistry. E. Arnold. Pub.

3

Cytology

INTRODUCTION

The development of genetics took a major step forward by accepting the notion that the genes, as characterized by Mendel are parts of specific cellular structures, the chromosomes. This unites the disciplines of genetics and cytology and correlates the results of breeding experiments with the behaviour of structures seen under the microscope. This fusion of genetics and cytology has important applications in medical genetics, agricultural genetics and evolutionary genetics.

The most prominent components of the cell nuclei are the chromosomes which have a constancy in the number from cell to cell within an organism and from organism to organism within any one species and from generation to generation within that species. Observing the chromosomal behaviour under the microscope during mitosis and meiosis will provide an understanding of the method of maintaining the chromosome number with respect to a species.

Modern instrumentation comprising computer aided high speed scanning devices has made it possible for the automatic identification of the photographic negatives of chromosomes. At the level of light microscopy, however, new techniques being discovered from time to time have helped in the clarification of chromosome details as to their structure. Use of special chemicals prior to fixation has been responsible for a clear understanding of the structure of the different parts of the chromosomes including the centromere. Pre-treatment for the study of chromosomes is carried out for clearing the cytoplasm, for softening and separation of the middle lamella and also for bringing out a good spread of chromosomes. Chemicals used for pre-treatment include normal HCl, enzyme preparations etc. The pre-treatment also secures a high frequency of metaphase stages through spindle inhibition. Chemicals such as colchicine, para-dichloro benzene, 8-hydroxy-quinoline bring about this effect known as C-mitosis (described later).

The tissues are fixed by selective chemicals. The fixation brings about the killing of the tissue without causing any distortion of the components. The tissues or their components are fixed selectively at a particular stage only. Special methods are used to study the phospholipid components of the chromosome or for the nucleoprotein precipitation. Freeze-drying methods are also used. All fixing chemicals are lethal and penetrate the tissue rapidly to kill it instantaneously so that the divisional figures are arrested at their respective stages enabling their study at varied stages. The fixative also prevents bacterial decomposition of the tissue. A good fixative must not interfere with the stains used for study. Non-metallic fixatives (ethanol, methanol, acetic acid, propionic acid, chloroform etc) are commonly used and metallic fixatives are used for special studies (osmium tetroxide etc). The most commonly used fixative used for light microscopic studies is the Carnoy's fixative (given at the end of the chapter).

Studies involving critical observations prefer the method of squashes and smears. In squashes, special treatments are needed to dissolve the middle lamella of cells. The material is then squashed after staining. This is suitable for the study of chromosome behaviour during mitosis in root-tip cells. In smears, the cells are spread directly over the slide and stained. This method is preferred for the study of meiosis in microspore mother cells of the anther. Staining is a must for study under a light microscope and light microscopy is preferred in most moderately equipped laboratories. Several different types of cytological stains are available. They can be classified as vital and non-vital (used for the study of living and killed tissue respectively). Stains can be acidic or basic and are chosen according to the need. Basic dyes are used for chromosome staining since chromatin is strongly acidic (acetocarmine) and mordants like metallic salts (e.g. iron) enhance the staining intensity. Feulgen staining also involves a procedure of acid hydrolysis which also has a mordanting effect. Certain amphoteric stains are also used (orcein). In this chapter, cytological methods involving the use of carmine, orcein and Feulgen are described.

CYTOLOGICAL PROCEDURES

Mitosis

Mitosis is the division of somatic cells. This produces a number of genetically identical cells from a single progenitor cell. Each single mitosis is associated with a single cell division that produces two genetically identical daughter cells. The cell cycle (Fig-19) can be divided into M (mitosis), S (DNA synthesis phase), G1 (gap 1) And G2 (gap 2) phases. Mitosis is the shortest period (about 5 - 10% of the total time taken by the cell cycle).

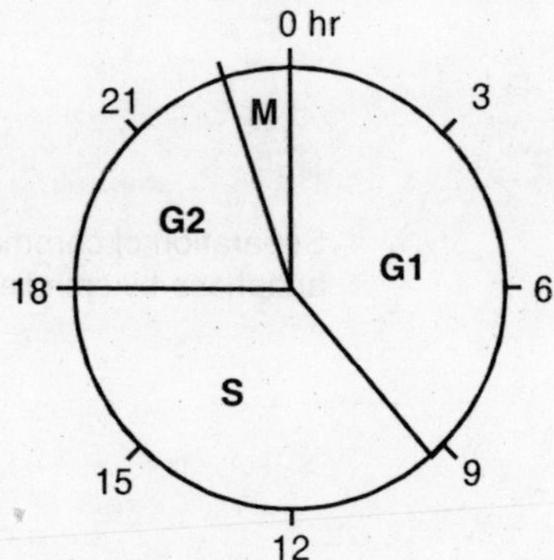

Fig. 19. Cell cycle

The rest constitutes the interphase. During mitosis, each chromosome in the nucleus duplicates longitudinally and this double structure splits to become two daughter chromosomes, each going to a different daughter cell. Mitosis generally comprises of the following stages (Fig-20).

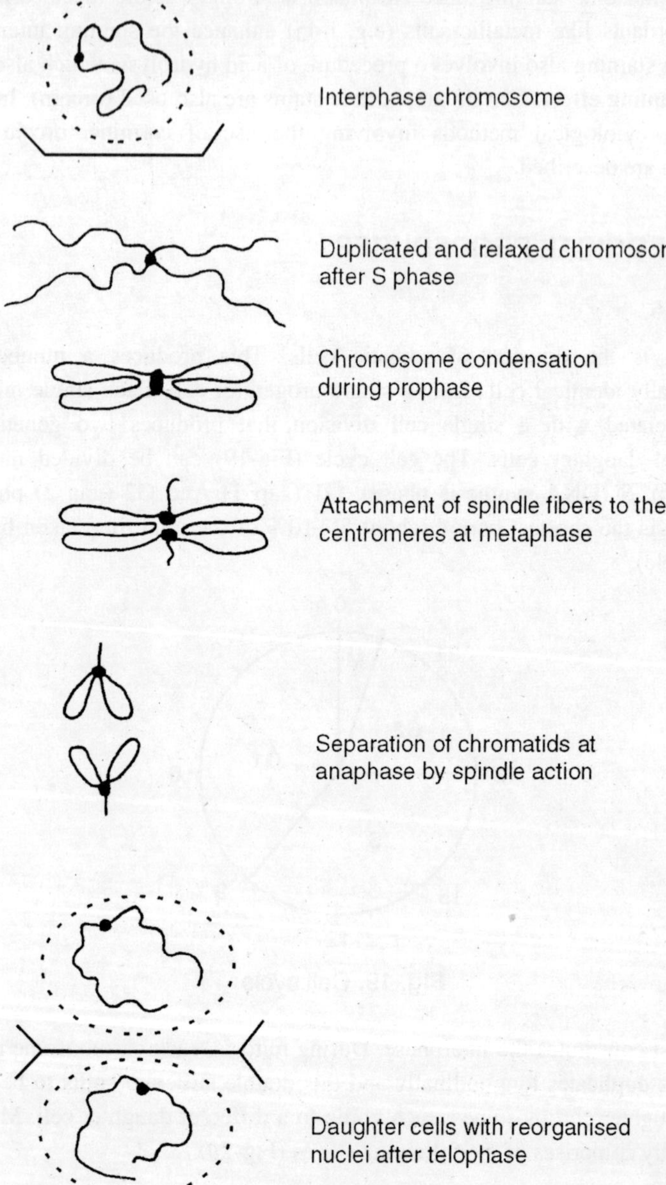

Fig-20. Mitosis

Prophase

The chromosomes become short and thick due to condensation and are comprised of two chromatids joined by a centromere. The nucleoli disappear and nuclear membrane breaks down.

Metaphase

The chromosomes get arranged on the equatorial plane with the chromatids of a chromosome facing both poles and the spindle fibres get attached to the centromeres.

Anaphase

The spindle fibres contract, thus dividing the centromere into two and pulling the chromatids to either pole.

Telophase

The separated chromatids arrange themselves and start recondensation followed by the appearance of nuclear membrane and nucleoli. The cytoplasm will also be divided into two and a new cell membrane will be laid down to separate the two daughter cells.

Mitosis can be studied in root tips of Allium cepa (onion), Allium sativum (garlic) and Vicia faba.

Study of mitosis in onion root tips

The process of mitosis can be studied in the dividing cells of the meristems. The root meristem of onion is a very convenient material for the study.

Requirements

- Onions
- Slides and coverslips
- Wide-mouthed glass containers
- Needles and brushes
- Filter paper and blotting paper
- Glacial acetic acid and 45% (v/v) acetic acid
- 1 N HCl

Grow fresh roots from onion bulbs.

Cut the roots when 1 - 2 cm long and immerse them in Carnoy's - I fixative.

Transfer to 70% ethanol and store at 4°C.

To prepare a squash for mitotic study, place a few roots in a watch glass containing 9 parts of 2% aceto-orcein and 1 part of 1N HCl.

Warm gently (do not boil) and leave covered for 15 min.

Pick up a single root, place on a slide and cut the root tip. Retain the root tip and discard the rest.

Squash the root tip in a drop of 45% acetic acid by gently pressing with a needle

Mount the preparation with a coverslip in a drop of 45% acetic acid and seal with Vaseline

Observe under the microscope. Tap over folds of blotting paper if the cells are not spread well

↓

Seal with the temporary seal and observe the small square cells (meristematic cells) for mitotic divisions.

Fig 21: Flow chart for processing of root tips to study mitosis.

- Ethanol
- Carnoy's I fixative: (Given at the end of the chapter)
- 2% aceto-orcein stain: (Given at the end of the chapter)
- Temporary seal: (Given at the end of the chapter)

Procedure

Fixation: Scrape the base of onions and stand them (with the help of pierced toothpicks) in water for 2-3 days. Roots start growing. Cut the root tips after they grow to a length of 1 cm and immerse them in the fixative solution. Transfer the root tips to 70% ethanol and store in the refrigerator at 4^0 C to use as and when required.

Staining and squash preparation: Place the root tips (that have been fixed and stored) in a watch glass. Add 9 parts of 2% aceto-orcein stain and 1 part of 1N HCl. Heat gently for 5 - 10 seconds three to four times. Leave covered for 10 - 15 min.

Transfer a root tip to filter paper to remove the acid mixture and place it on a clean slide. Cut off the very short meristematic end (which has taken up more stain) and discard the rest. Add 2% aceto-orcein and add the cover slip. Hold one edge of the cover slip down so it cannot move, tap the cover slip with a rubber tipped pencil 2 or more times. Then place a folded filter paper over the cover slip and apply gentle pressure with the thumb. Roll the thumb over the folded filter paper. This helps in separating the cells in division from the rest of the tissue. If needed, tap a little over the cover slip a little more. Observe the different stages of mitosis under the microscope under lower & higher magnification. Seal the slides with temporary seal and label them. The above flow chart (Fig-21) is provided for convenience.

Study of C - mitosis (colchicine treated root tips)

Colchicine has a dramatic effect on the chromosomes. It shortens the chromosomes and prevents spindle formation. This results in clearly spread-out chromosomes, which can easily be counted, whereas non-colchicine treated cells show side views of metaphase, which are not usually clear. Other pre-treatment chemicals that produce a similar effect include para dichloro benzene and 8 - hydroxy quinoline.

Requirements

- Onion bulbs with healthy growing roots
- 0.1% (w/v) colchicine
- Carnoy's - I fixative (given at the end of the chapter).
- 2% aceto-orcein stain (given at the end of the chapter).
- Temporary seal (given at the end of the chapter).

Procedure

Colchicine treatment: Stand the onion bulbs with healthy growing roots, (about 1 cm long) in 0.1% colchicine for 4 hours in a refrigerator. Wash them thoroughly with dH_2O and fix them in Carnoy's-I fixative.

Method: Proceed as described above to prepare the squashes of root tips.

Study of karyotype of *Allium cepa*

Karyotype is the presentation of chromosome features in a cell nucleus. The position of the centromere determines the shape of the chromosomes. The centromere is known as the primary constriction. The shape of the chromosomes ranges from *metacentric* (when both the chromatids are of equal size), *sub-metacentric* (when one chromatid is slightly longer than the other), *acrocentric* (when one chromosome is very tiny compared to the other) and *telocentric* (when the centromere is located at the tip of the chromosome). Besides these constrictions, there are certain secondary constrictions with a satellite (knob like structure) at its end. Such a chromosome is called as a *sat chromosome.* Chromosomes of a species have a fixed size, shape, number etc. These specific attributes are referred to as the karyotype. The karyotype is generally studied by scanning the c-metaphase plates.

Requirements:

- Freshly grown root tips
- 2% aceto-orcein (given at the end of the chapter)
- Carnoy's-I fixative (given at the end of the chapter)
- % (w/v) colchicine
- Temporary seal material (given at the end of the chapter)

Procedure

Pre-treatment: Pre-treat the root tips with 0.1% colchicine solution for four hours by placing them in a refrigerator. Wash the roots with dH_2O and fix them in Carnoy's-1 fixative.

Method: Proceed as stated above and prepare the aceto-orcein squashes. Study the cells with scattered metaphase plates and record the number of chromosomes, shapes of chromosomes with respect to the position of the centromere, sizes of different chromosomes and number of satellite chromosomes.

Effects of chemical mutagens on somatic chromosomes

Chemical mutagens produce anomalies in somatic cells. These include laggards, anaphasic bridges, breaks and other abnormalities.

Requirements

- Healthy growing roots
- Ethyl methane sulphonate (chemical mutagen)(w/v): 0.05%, 0.1% and 0.2%.
- Carnoy's-I fixative (given at the end of the chapter)
- Aceto-orcein stain (given at the end of the chapter)
- Temporary seal (given at the end of the chapter)

Procedure

Treat the healthy growing roots with the different concentrations of ethyl methane sulphonate for two hours and wash them with dH_2O before fixing them in the fixative. Maintain appropriate controls (treat with dH_2O). Prepare the squash as described earlier with aceto-orcein and observe under the microscope. Score the number of cells undergoing abnormal mitosis and also record the different types of abnormalities like anaphasic bridges, laggards, breaks etc.

Localization of DNA by Feulgen method

Feulgen reaction is a specific test developed to locate DNA by a reaction between leucobasic fuchsin and the aldehyde group in pentose sugar of DNA. The intensity of the colour may be used to quantitate the DNA.

Requirements:

- Root tips fixed in Carnoy's-I fixative
- Feulgen reagent (given at the end of the chapter)
- 45%(v/v) acetic acid
- 5N HCl
- Temporary seal (given at the end of the chapter)

Procedure

Hydrolyze the root tips in 1N HCl at 60° C for 20 min. Alternatively, hydrolyze the root tips in 5N HCl at room temperature for 20 min. Wash them thoroughly with dH_2O and transfer to leucobasic fuchsin stain taken in a dark bottle. Leave in the dark for 3 hours. Mount one root tip in 45% acetic acid and seal with the temporary seal material. Observe the preparation under the microscope.

Orcein-banding of somatic chromosomes

The orcein-banding of chromosomes reveals dark banding around the centromere comprising constitutive heterochromatin.

Requirements

- Healthy growing onion roots
- 1% (w/v) glucose
- 0.2% (w/v) colchicine
- 1 N HCl
- 2% aceto-carmine stain (given at the end of the chapter)
- Carnoy's-I fixative (given at the end of the chapter)

Procedure

Treat the roots with 1% glucose for 30 min and later with 0.2% colchicine for 4 hrs in the refrigerator. Transfer the roots to a 1:1 mixture of 1% glucose and 0.2% colchicine and incubate for 1 hr. Fix in Carnoy's-I fixative for 24 hrs. Prepare the squash as described earlier. Observe the dark bands around the centromeres.

MEIOSIS

Meiosis is a reduction division, which results in four daughter cells (gametes), each containing half the number of chromosomes (n). These gametes fuse during fertilization restoring the original diploid chromosome number (2n) in the zygote. A premeiotic S-phase will precede the meiosis where the required DNA will be synthesized. Meiosis consists of two cell divisions: Meiosis I and Meiosis II (Fig-22).

Meiosis I is divided into a lengthy prophase (called prophase I) which is comprised of leptotene, zygotene, pachytene, diplotene and diakinesis. The homologous chromosomes approach each other during prophase I and synapse. Each chromosome is made up of two chromatids. Crossing overs occur during synapsis at the region of chiasmata between the non-sister chromatids of the homologous pair of chromosomes, resulting in exchange of chromosome material called genetic recombination. Towards the end of prophase and during metaphase, the chromosomes contract to become very short and stout. The homologous chromosomes separate out by spindle action at Anaphase I and approach separate poles in Telophase I.

Meiosis II starts with prophase II, where only half the number of chromosomes would be present. The metaphase II, anaphase II and telophase II resemble the mitosis. The chromatids of each chromosome separate out due to spindle action and reach opposite poles in both the daughter cells, thus resulting in a quartet or a tetrad. Each daughter cell would therefore contain only half the somatic number of chromosomes (n).

Meiosis, in microspore (pollen) mother cell (PMC) in pollen sacs leading to formation of microspores or pollen grains is known as microsporogenesis. Meiosis, in megaspore mother cell in ovule leading to the formation of megaspore is known as megasporogenesis. In animals, sperms and ovum formation resulting due to meiosis are known as spermatogenesis and oogenesis respectively.

Pollen mother cells are the preferred material for study of meiosis. They are large spherical thick walled cells and present a clear view of the chromosomes.

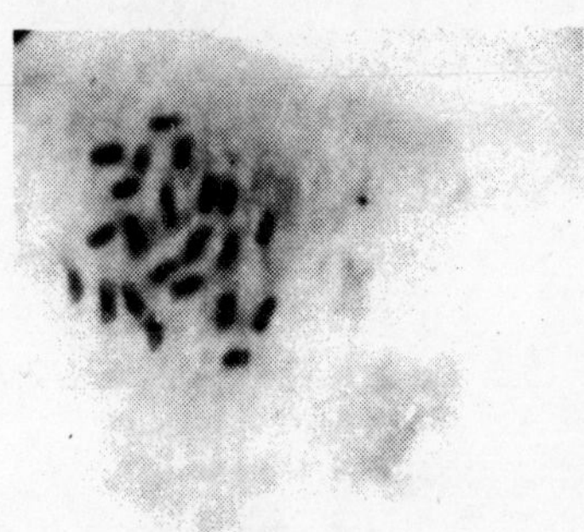

Fig. 22: Meiosis.

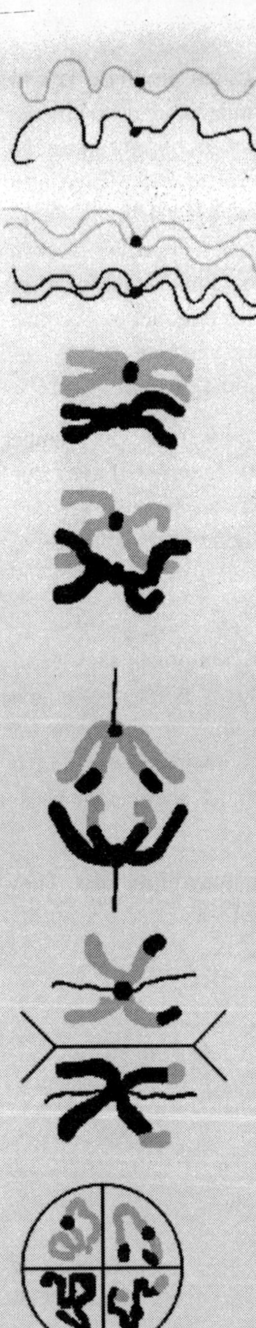

Relaxed homologous chromosomes in a germ cell before meiosis

Chromosomal duplication at prophase-I (leptonema)

Pairing of homologous chromosomes and condensation at prophase-I (zygonema)

Chiasmata between homologous chromosomes at prophase-I (pachynema) and exchange of chromosomal material (Crossing over)

Separation of the homologous chromosomes at anaphase-I. Note that each pair has a recombinant chromatid

Reorganized chromosomes set for the II meiotic division (at metaphase-II)

Separated chromatids get reorganized after meiosis-II into tetrads (haploid cells)

Study of meiosis in maize sporocytes

The most suitable plant material for the study of meiosis is the maize microsporocyte material. The appropriate stage for the study of meiosis is about 5 to 10 days before the tassel tip begins to appear (approximately the 21st day from the seedling appearance).

Requirements

- Maize plants in field (20-21 days old)
- Carnoy's -II fixative (given at the end of the chapter)
- 2% aceto-carmine stain (given at the end of the chapter)
- Temporary seal (given at the end of the chapter)

Procedure

Fixation: Fix the entire inflorescence in the fixative (Carnoy's II fixative). The spikelets are progressively younger towards the tip. Several trials may be required for the selection of the right type of sporocytes that are in the dividing stages. Larger ones may have pollen grains and are therefore not useful. So are the very young ones. Store the material at 4^0C in 70% alcohol after incubating in the fixative for 36-48 hrs.

Staining and smear preparation: Dissect out the anthers from the florets in a drop of 2% aceto-carmine stain. Use rusty needles for dissecting the florets and stirring the aceto-carmine stain to incorporate iron into the stain. Iron helps in good staining of the chromosomes. Cut the anthers with the help of needles and tap the anther pieces to extrude the sporocytes. Stir the drop well with the needle. Pick up the shells of anthers with two needle tips and discard them. There will approximately be 750 to 1000 microspore (pollen) mother cells in a single anther. Place the cover slip carefully and see that the drop of stain fills up well. Do not use excessive stain since the cells will move out of the cover slip area. Place a folded filter paper over the slide and press gently before applying the temporary seal. View the slide under highest magnification of the microscope. A flow chart (Fig-23) is provided.

Collect the maize inflorescences before they emerge out of the flag leaf (before the pollen grains are formed) and fix them in Carnoy's - II fixative.

↓

Transfer them to 70% ethanol and store at 4°C.

↓

Dissect a single flower on a clean slide and isolate the anthers. Discard the debri.

↓

Put 2 drops of 2% aceto-carmine on the anthers and heat gently over a spirit lamp. Turn them around with a rusted needle and squash them well.

↓

Lower a coverslip over the material (add some more stain if needed) and place a piece of blotting paper over the coverslip and tap firmly with a blunt needle.

↓

Seal the coverslip with temporary seal and observe the preparation under a microscope. Scan the large circular cells (micro spore mother cells) for various stages of meiosis. (The tiny square anther wall cells should be ignored).

Fig 23: Flow chart for processing of maize flower buds to study meiosis.

Study of translocation heterozygosity at meiosis in plants

Reciprocal translocations at meiosis result in quadrivalents and rings. This can be studied in flower buds of *Tradescantia* or *Rhoeo discolor*.

Requirements

- Flower buds of Tradescantia / Rhoeo discolor
- Carnoy's-II fixative (given at the end of the chapter)
- 2% aceto-carmine (given at the end of the chapter)

- Temporary seal (given at the end of the chapter)

Procedure

Fix the flower buds in the fixative and incubate them for 2 hrs. Prepare the smear as described earlier and observe the preparation under the microscope for the rings of reciprocal translocations.

Cytology of cell suspension culture

Cell division (mitosis) in the *in vitro* cultured suspension cells can be studied by special pre-treatment prior to the fixation.

Requirements

- Plant suspension culture (3 days old culture)
- Carnoy's-I fixative (given at the end of the chapter)
- M sodium acetate (pH 6.5)
- 45% (v/v) acetic acid
- 0.5% cellulase enzyme
- 0.5% pectinase enzyme
- 2% aceto-carmine stain (given at the end of the chapter)
- Temporary seal (given at the end of the chapter)

Procedure:

Collect the three-day-old suspension culture cells by centrifugation and transfer them to 0.2% colchicine for 8 - 10 hrs. Fix the cells in Carnoy's-I fixative for 24 hrs. Wash the cells with 0.1M sodium acetate and digest the cell walls with the cellulase and pectinase enzymes. Rinse twice with 45% acetic acid and store in 45% acetic acid at 4^0C. Apply a small drop of cells to a slide and add a drop of 2% aceto-carmine. Mix with a rusted needle and place a cover slip. Press the cover slip gently over folds of blotting paper and seal with the temporary seal. Observe under the microscope.

PREPARATION OF FIXATIVES

1. **Carnoy's-I fixative:** Mix three parts of absolute alcohol to one part of glacial acetic acid.

2. **Carnoy's-II fixative:** Mix six parts of absolute alcohol to 3 parts of chloroform and one part of glacial acetic acid.

PREPARATION OF STAINS

1. **2% aceto-orcein (w/v):** Dissolve 2 GM orcein in 100 ml of 45% acetic acid and reflux for about 30 **min** by gentle boiling. Cool and filter. Store in a refirgerator in a brown dropper bottle.

2. **2% aceto-carmine (w/v):** Dissolve 2 GM carmine in 100 ml of 45% acetic acid and reflux for about 2 hours by gentle boiling. Then store in a refrigerator in a brown dropper bottle.

3. **Basic fuchsin and preparation of Feulgen reagent:** Basic fuchsin is a mixture of p-rosaniline chloride, rosaniline chloride and new magenta. This is a magenta coloured dye. It is soluble in water and alcohol. When an aqueous solution of basic fuchsin is treated with sulphurous acid, a colourless leuco-basic fuchsin-sulphurous acid is obtained called as Schiff's reagent or Feulgen-reagent. This stain is specific for staining DNA in the chromosome.

Preparation of Feulgen reagent:

Requirements

- Basic fuchsin 0.5 g
- 1 N HCl 10 ml
- Potassium metabisulphate 0.5 g
- Activated charcoal 3.5 g
- Distilled water 100 ml
- Whatman No 1 filter paper

Procedure

Dissolve 0.5 g of basic fuchsin in 100 ml of boiling water. Cool to 60^0C. Filter through Whatman No 1 filter paper. Then add 10 ml of 1 N HCl and 0.5 g potassium metabisulphate. Add charcoal powder after 24 hrs, which will decolourize the solution (leucobase). Store in a dark bottle in a dark place.

PREPARATION OF TEMPORARY SEAL

Mix equal proportions of paraffin and vaseline and store in a glass container.

REFERENCES

1. Avers. C. (1982) Cell Biology. 2nd edition. D. Van. Nostrand. New York.
2. Avers. C. (1982) Genetics. 2nd edition. D. Van. Nostrand. New York.
3. Griffiths. A. J. F., Miller. J. H., Suzuki. D. T., Lewontin and Gelbart. W. M (1993)
4. An introduction to Genetic Analysis. 5th edition. Freeman and Company, New York.
5. Sharma. A. K. and Sharma. A (1990) Chromosome techniques: Theory and practice, Third edition. Butterworth and Co. Publ. Ltd.

4

Plant Genetics

Johann Gregor Mendel laid the foundations of modern genetics with the publication of his pioneering work on peas in 1866. He proposed a theory of particulate inheritance. However, it was appreciated only after the rediscovery of the laws of heredity in 1900. The knowledge of genetics in the past century has grown to a stage where the genetic material is being exchanged between unrelated organisms at present. Certain basic Mendelian principles are presented here.

PRINCIPLE OF SEGREGATION

A genetic factor (gene) specific for a character is passed on from one generation to the next as a unit (without any mixing of units). Only one allelic form of a gene is transmitted through a gamete to the offspring. There are two alleles of each gene. Any two alleles are situated on the homologous chromosomes (one allele on each homologous chromosome). Alleles can be dominant or recessive. An individual can have both dominant alleles or both recessive alleles (called homozygotes for the trait) and can be recognized by the appearance (phenotype). However in an individual possessing one dominant and recessive allele, the appearance (phenotype) will be of the dominant type (there are intermediate forms sometimes). But as per the alleles, the individual is a heterozygote (according to the genotype). Hence, the principle of segregation supports the fact that the two alleles segregate during gamete formation and will enter different gametes during meiosis.

PRINCIPLE OF INDEPENDENT ASSORTMENT

The segregation of one factor (gene) pair occurs independently of any other factor pair [i.e. if these genes are located on non-homologous chromosomes only]. The segregation and assortment will be random and results in two parental types and two recombinants.

Example

If there are three pairs of homologous chromosomes carrying three different genes **A/a, B/b,** and **C/c,** how many different assortments or combinations are expected in the gametes?

Solution: Use the following forking system:

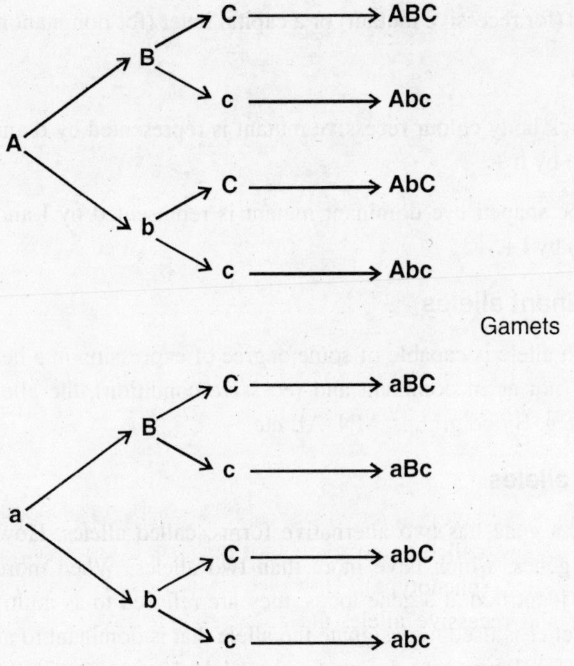

TERMINOLOGY IN GENETICS

Phenotype

Measurable, visible, distinctive traits of an organism e.g. colour of flower, or hair texture. Blood groups (require tests) are also phenotypes.

Genotype

All of the genes present in an individual constitute its genotype. Generally when a single locus is studied, the one gene status is represented as genotype **AA** and when two loci are studied, the two-gene status is represented as genotype **AABB.**

Dominant and recessive alleles

The allele that can express itself in the heterozygote is called the dominant factor and is represented by the capital letter whereas the other allele is the recessive factor and is represented by the lower case letter. However in the study of normal and mutant types, the normal allele is represented by **+** and its phenotype is called **wild type.** The **mutant type** is represented by a lower case letter (for recessive mutant) or a capital letter (for dominant mutant).

Example

- A black body colour recessive mutant is represented by **b** and wild type (grey) by **b +.**
- A lobe shaped eye dominant mutant is represented by **l** and wild type (oval) by **l** +.

Co-dominant alleles

When each allele is capable of some degree of expression in a heterozygous condition (not as in dominant and recessive condition), the alleles are co-dominant e.g. Blood groups, MN, AB etc.

Multiple alleles

Generally, a gene has two alternative forms, called alleles. However there are many genes, which have more than two alleles. When more than two alleles are identified at a gene locus, they are referred to as multiple alleles. A capital letter is used to designate the allele that is dominant to all others in the series. The lower case designates the allele that is recessive to all others in the series. All other intermediate alleles are assigned the lower case letter with some suitable superscript, e.g. red or wild type eyes in drosophila is represented by **w** or $\mathbf{w}^{+}$ and the recessive white eye by **w**. The $\mathbf{w}^{+}$ and **w** are thus completely dominant and completely recessive respectively. All other intermediate possibilities are represented according to the intensity of the pigment produced. $\mathbf{w}^{\mathbf{co}}$ (cora), $\mathbf{w}^{\mathbf{bl}}$ (blood), $\mathbf{w}^{\mathbf{e}}$ (eosin), $\mathbf{w}^{\mathbf{ch}}$ (cherry), $\mathbf{w}^{\mathbf{a}}$ (honey) etc.

Incomplete dominance

The heterozygous genotype gives rise to a phenotype distinctly different from either of the homozygous genotypes. It is intermediate in character produced by a sort of *blending*. The alleles however maintain their individual identities and segregate from each other in the formation of gametes.

Lethal alleles

Some genes express by causing the death of the individual prior to maturity or even in the prenatal stage. These are the lethal genes. These are dominant lethal (the individual dies at infancy) or recessive where they are lethal in homozygous condition.

SINGLE GENE INHERITANCE

For every single gene, a pair of alleles govern the trait (e.g. tall (**T**) and dwarf (**t**) traits in pea). There are 6 types of crosses possible among the three genotypes of **TT, tt** and **Tt.** They are given below:

Matings	Expected F1 ratios	
	Genotypes	Phenotypes
TT x TT	All TT	All tall
TT x Tt	½ TT : ½ Tt	All tall
TT x tt	All Tt	All tal
Tt x Tt	¼ TT : ½ Tt : ¼ tt	¾ tall : ¼ dwarf
Tt x tt	½ Tt : ½ tt	½ tall : ½ dwarf
tt x tt	All tt	All dwarf

F1 and F2 generations

The parental generation is symbolized by P **and** the first filial generation of offspring is symbolized as F1 (monohybrid since a single gene is considered). Unless otherwise specified in each case, the second filial generation (F2) is produced by crossing F_1 individuals among themselves randomly. Self-fertilizing plants can also be manually cross-pollinated in the first parental cross. The F_1 and F_2 are produced as follows:

TT	×	**tt**	**P**
Tall		White	
(T)		(t)	**gametes**
	Tt		.... F_1

Tt	×	**Tt**	
(T) (t)		(T) (t)	**gemetes**
¼ TT : ½ Tt : ¼ tt			. ... F_2

Genotypic ratio = 1:2:1; Phenotypic ratio = 3:1 [tall : dwarf]

Test cross

This is used to distinguish between a homozygous dominant genotype and heterozygous genotype (since both have the same phenotype). A homozygous recessive phenotype is crossed with such a phenotype.

Tall [?]	**X**	**Dwarf [tt]**
	All tall (Tt) ....Offspring	

All offspring are black i.e. **Tt.** This means that the tall parent has produced only one kind of gamete **(T).** Hence it is a homozygous dominant genotype **TT.**

Tall [?]	**X**	**Dwarf [tt]**
½ (Tt)	and	½ (tt) ....Offspring
Tall		Dwarf

This means that the tall parent has produced two types of gametes and hence is heterozygous **Tt.**

Back Cross

If F_1 progeny are mated to one of their parents, it is termed back cross.

TT × **tt** **p**

Tt .... F_1

Tt × **TT** **Back cross**

(T) (t) (T) gametes

½ TT and ½ Tt (All tall) .... Back cross progeny

Inheritance of two or more genes

Two or more traits each controlled by a different pair of independently assorting autosomal genes can be studied simultaneously and referred to as a dihybrid genotype or trihybrid genotype as the case may be.

Dihybrid genotypes that are heterozygous at both loci form four genetically different gametes and the F_1 and F_2 are produced in the same pattern as seen in case of monohybrid crosses.

Four types of gametes i.e. (**BL**), (**Bl**), (**bL**) and (**bl**) are produced by a dihybrid black, short haired guinea pig (**Bb Ll**). Other genotypes have different gametic out put when compared:

BB LL	All BL
BB Ll	½ BL : ½ Bl
BB ll	All Bl
Bb LL	½ BL : ½ bL
Bb Ll	¼ Bl : ¼ Bl : ¼ bL : ¼ bl
Bb ll	½ Bl : ½ bl
bb LL	All bL
bb Ll	½ bL : ½ bl
bb ll	All bl

Dihybrid F2 ratio

When two genotypes are crossed, **BB LL x bb ll**, the F1 will be a dihybrid Bb Ll. Self-fertilization of F_1 will result in F_2 progeny, which will be in the ratio of 9 : 3 : 3 : 1 (in contrast to 3 : 1 of monohybrid cross).

Bb Ll × Bb Ll

(BL)(Bl) (bL) (bl)..........gametes (both parents produce identical gametes)

	BL	Bl	bL	bl
BL	BBLL	BBLl	BbLL	BbLl
Bl	BBLl	Bbll	BbLl	Bbll
BL	BbLL	BbLl	bbLL	bbLl
Bl	BbLl	BbLl	bbLl	bbli

Genotypic ratio :- 1:2:1:2:4:2:1:2:1
Phenotypic ratio:- 9:3:3:1

Genotypes	*Phenotype*
1/16 BBLL	
2/16 BBLl	Black & Short
2/16 BbLL	**9/16**
4/16 BbLl	
1/16 Bbll	Black & Long
2/16 Bbll	**3/16**
1/16 bbLL	White & Short
2/16 bbLl	**3/16**
1/16 bbll	White & Long
	1/16

Test crosses (between an incompletely known genotype and a homozygous recessive genotype) will reveal the number of different gametes formed by the parental genotype. A monohybrid test cross gives 1:1 phenotypic ratio whereas a dihybrid test cross gives 1:1:1:1 ratio indicating that there are two pairs of genes involved.

PREDICTING OUT COME OF CROSSES AND PROBABILITY FORMULATIONS

If a monohybrid cross is observed, where fertilization occur at random and the chance that any given egg will be fertilized by any given sperm is the product of t heir separate probabilities. Hence in a cross: **Aa x Aa,**

Eggs .... (A) and (a)

Gametes

Sperms .... (A) and (a)

egg A fertilized by sperm A = 1/2 × 1/2 or 1/4 AA progeny

egg A fertilized by sperm a = 1/2 × 1/2 or 1/4 Aa progeny

egg a fertilized by sperm A = 1/2 1 1/2 or 1/4 Aa progeny i.e 1/2 Aa progeny

egg a fertilized by sperm a = 1/2 × 1/2 or 1/4 aa progeny

= 1 : 2 : 1 ratio (AA : Aa : aa).

Probability of obtaining of heterozygote from such parental types is 1/2 and probability of obtaining a homozygote **AA** or aa is 1/4. Probability of producing an individual with a dominant trait is 3/4. Probability of producing an individual with a homozygous recessive phenotype is 1/4. This can be deduced from the rule of probability that the chance of two independent events occurring simultaneously is the product of their separate probabilities.

Using these probability formulations, we can predict the outcome of genetic crosses of more than one trait also.

CHI - SQUARE TEST

The Chi-Square test ($\mathbf{X^2}$) is a simple statistical test, which is used to determine whether a given set of data fits a particular ratio (e.g. 3:1 or 9:3:3:1 genetic ratios).

$X^2 = \Sigma\,(D^2 - E)$ *where* Σ = stands for sum of

D = deviation from expected ratio i.e.

D = O - E *where* O = observed value ,E = expected value.

The method of calculation of Chi - Square and problems on Chi - Square are included towards the end of this chapter.

EXERCISES AND PROBLEMS

Exercises: Study of monohybrid cross by using maize cobs

Monohybrid crosses can be studied from maize cobs (Fig-24) having purple (coloured, due to anthocyanin pigment) and yellow (colourless) grains which could be picked up from maize fields (they should belong to F_2 generation).

Requirements: Maize cobs with coloured and colourless grains.

Procedure: Count the total number of rows of grains on the cob. Count the number of coloured and colourless grains in each row. Calculate the total number of coloured and colourless grains. Then calculate the ratio between colourless and coloured grains. eg :

If the number of coloured grains is 145 and the number of colourless grains is 46, then the monohybrid ratio for grain colour is 3.1 : 1, which is approximately equivalent to Mendel's monohybrid ratio of 3 : 1. The colour grain type is dominant over colourless grain type. The inheritance of this grain colour represents complete dominance.

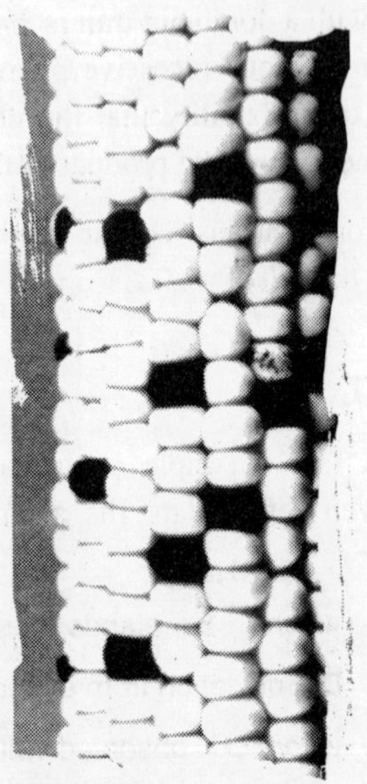

Fig. 24. Maize cob

Problems

Problem 1

Heterozygous black **(Bb)** mice are mated to homozygous recessive **(bb)** whites. What phenotypic and genotypic ratios are expected from back crossing the black F_1 progeny to (a) the black parent? (b) the white parent?

Solution: a). Genotypic ratio = 1/4 BB : 1/2 Bb : 1/4 bb.

Phenotypic ratio = 3/4 black : 1/4 white.

b) Genotypic ratio = 1/2 Bb : 1/2 bb.

Phenotypic ratio = 1/2 Bb black : 1/2 bb white

Problem 2

Two types of phenotypes are available in tomatoes, yellow or red fruit. Plants of these two phenotypes were crossed as follows:

Parents:	Progeny:
red x red	60 red
red x red	45 red, 15 yellow
red x yellow	60 red
yellow x yellow	65 yellow
red x yellow	37 red, 38 yellow

(a) What phenotype is dominant?

(b) What are the genotypes of the parents and progeny in each cross?

Solution

(a) The phenotype red is dominant since the second cross resulted in a 3 : 1 phenotypic ratio showing that the parents are heterozygous red the third cross resulted in red progeny, which is a F1 hybrid and the fifth cross resulted in a test cross ratio of 1:1 meaning that the parents are heterozygous red and homozygous recessive yellow.

(b)

Genotypes of parents:	Genotypes of progeny:
RR × RR or RR x Rr	all RR or 1 RR : 1 Rr
Rr × Rr	1 RR : 2 Rr : 1 rr
RR × rr	all Rr
rr × rr	all rr
Rr × rr	1 Rr : 1 rr

Problem 3

A cross between two 4 'o' clock plants produced 80 plants with pink flowers, 36 with red flowers and 34 with white flowers.

(a) What is the genotype and phenotype of each parent?

(b) What phenotypes would you expect and in what proportions among the progeny of the following crosses:

(1) pink × pink, (2) red × red, (3) red × white, (4) pink × white.

Solution

(a) The phenotype of the two parents is pink and genotypes of both are Rr.

(b) 1. Phenotypic ratio = 1/4 red : 1/2 pink : 1/4 white

2. RR = All

3. Rr = All pink

4. Rr = rr

1 *pink* : 1 *white* (Incompletely dominant pair of alleles)

Problem 4

In fruit-fly, scarlet eye colour is produced in homozygous recessive flies while normal flies have red eyes.

(a) If red eyed females are crossed with scarlet eyed males, what will be the eye colour of the offspring?

(b) If males and females of the F_1 progeny are interbred, what phenotypes can we expect in the F_2 and in what proportions?

(c) If an F_1 female is crossed to a scarlet-eyed male, what kinds of progeny will be produced and in what proportions?

Solution

(a) Rr (Red eyed) ...... F_1

(b) Phenotypic ratio = 3 red : 1 scarlet.

(c) 1 red : 1 scarlet

Problem 5

A man and his wife are both tasters of Phenyl Thiocarbamide **(PTC).** Out of four children, two are non-tasters. What are the parental genotypes?

Solution

Tt × **Tt**., where **T** is the gene for tasters and **t** for non-tasters.

Problem 6

In mice, the dominant allele for coat colour is **C** and its recessive allele for albino is **c** and the dominant allele for hair type is **S** for straight hair and **s** for bent hair. Find out the following from the progeny of a dihybrid cross (**CcSs x CcSs**).

(a). The proportion of albinos.

(b). The proportion of mice with bent hair.

(c). The expected ratio of phenotypes in F_2.

Solution

(a). 1 /4.

(b). 1 /4.

(c). 9, coloured coat and straight hair : 3, coloured coat and bent hair : 3, albinos and straight hair : 1, albino and bent hair.

Problem 7

What is the proportion of homozygous offspring from the cross **AaBbCc x AaBbCc** ?

Solution: 1/8

GENETIC INTERACTION

Genetic interaction occurs whenever two or more genes specify enzymes that catalyze steps in a common pathway. If an essential substance is not produced it results in defective enzymes leading to an abnormal (mutant) phenotype. The mutant phenotype would be the normal wild type if one of the enzymes (e.g. in a total of two enzymes x and y, only x is defective) would not be defective. It means that the normal expression of y would depend only on the availability of x. This means that the expression of one gene depends on the expression of another gene due to gene interaction. The term epistasis (meaning suppression of the hypostatic gene) has developed a new meaning of interaction by mutual epistatic effect of both the genes on one another. Dominance involves intra-allelic gene suppression whereas epistasis involves inter-allelic gene suppression. The classical phenotypic ratio of 9 : 3 : 3 : 1 observed in the progeny of dihybrid parents becomes modified by epistasis into ratios that are various combinations of the 9 : 3 : 3 : 1 groupings and the number of phenotypes appearing in the offspring from dihybrid parents will be less than 4. They are:

Dominant epistasis: (Ratio 12 : 3 : 1)

Only when the epistatic locus has homozygous recessive alleles (**aa**) can the alleles of the hypostatic locus (**B or b**) be expressed.

Recessive epistasis: (Ratio 9 : 3 : 4)

Only if the dominant allele is present at the **A** locus can the alleles of the hypostatic **B** locus be expressed.

Duplicate genes with cumulative effect: (Ratio 9 : 6 : 1)

The dominant condition (either homozygous or hetrozygous) at either locus (but not both) produces the same phenotype.

Duplicate dominant genes: (Ratio 15 : 1)

The dominant alleles of both loci each produce the same phenotype (i.e. in either of the two genes, a dominant allele will produce the same phenotype) without cumulative effect except the all recessive genotype.

Duplicate recessive genes: (Ratio 9 : 7)

Both dominant alleles (of the two genes) when present together complement each other to produce a different phenotype.

Dominant and recessive interaction: (Ratio 13 : 3)

A dominant genotype at one locus (e.g. **A**) and the recessive genotype at the other **(bb)** produce the same phenotype i.e. **A-B-, A-bb** and **aa bb** will produce the same phenotype and **aa B** - produces the other.

Problems

Problem 1

Dominant epistatic gene **I-** is an inhibitor of pigment production in onion bulbs. It is epistatic over another locus, the genotype **iiR** - producing red bulbs and **iirr** producing yellow bulbs.

(a) A pure red strain is crossed to a pure white strain and produces an all white F_1 with 12/16 white, 3/16 red and 1/16 yellow. What were the genotypes of the parents?

(b) If yellow onions are crossed to a pure white strain of a genotype different from the parental type in part (a), what phenotypic ratio is expected in the F_1 and F2.

Solution

(a) A pure red strain can be **iiR** and a pure white strain can be **I--.** The F1 are all white **I-R-.** The F2 were in 12 : 3 : 1 ratio. Therefore the F1 was a hybrid IiRr and hence the parents were **IIrr** and **iiRR.**

b) F1 is the same as in part (a), hence will produce 12/16 white : 3/16 red :1/16 yellow.

Problem 2

In certain garden plants, the recessive genotype of one locus (**aa**) prevents the development of pigment in the flower, thus producing a white colour. In the presence of the dominant allele **A**, alleles at another locus may be expressed as follows: **C** - = red, **cc** = cream.

(a) When plants with cream flowers of the genotype Aa **cc** are crossed to those with flowers of the genotype **Aa Cc**, what phenotypic and genotypic proportions are expected in the progeny.

(b) When dihybrid plants (red flowered) are crossed together, what phenotypic ratio is expected among the progeny?

Solution

(a) Phenotypic ratio = 3/8 red : 3/8 cream : 1/4 white
Genotypic ratio = 1/8 AA Cc : 1/8 AA cc : 1/4 Aa Cc : 1/4 Aa cc : 1/8 aa Cc : 1/8 aacc.

(b) Phenotypic ratio = 9 red : 3 cream : 4 white

Problem 3

Summer squash produces fruits with three different shapes: disc shaped, elongated and sphere shaped. A pure disc-shaped variety was crossed to a pure elongated variety. The F_1 were all disc shaped. Among the F_2, what is the proportion of sphere shaped, elongated and disc shaped fruits. What is the type of interaction involved?

Solution

Phenotypic ratio: 9/16 Disc shaped : 6/16 Sphere shaped : 1/16 Elongated shaped. The type of genetic interaction is duplicate genes with cumulative effect.

Problem 4

The digitalis plant produces a seed capsule, whose shape is controlled by two independently assorting genes. When dihybrid plants were crossed, 5% of the progeny had ovoid shaped seed capsules. The other 95% of the progeny had triangular- shaped seed capsules, what phenotypic ratio is seen among the F_2 and what type of interaction of genes is operative?

Solution

All except **the** double recessive **(ttoo)** are of triangular shaped seed phenotype (15 : 1)

Problem 5

Coloured grains in maize are produced by the interaction of two dominant genes **A_1** and **A2.** All other genetic combinations of alleles produce colourless grains. Two pure colourless strains are crossed to produce an all coloured F_1. a) What were the genotypes of the parental strains and the F_1? b) What phenotypic proportions are expected among the F_2 and what is the genetic interaction involved?

Solution

(a) **A_1A_1 a_2 a_2 × a_1a_1 A_2 A_2** colourless grain (parents) and **A_1a_1 A_2a_2** coloured F_1 hybrid.

(b) Phenotypic ratio = 9 coloured grains : 7 colourless grains. This 9 : 7 ratio has been produced due to the duplicate recessive gene interaction.

Problem 6

In onion bulbs, white is epistatic to coloured ones. The dominant gene **I--** suppresses the gene for coloured bulbs **C.** All combinations except **iiC** - will produce white bulbs. A white variety homozygous for the dominant alleles is crossed with a double recessive white variety. What are the phenotypic F_1 and F2 ratios and what is the interaction operative?

Solution

Phenotypic ratio = 13/16 white : 3/16 coloured.

The dominant and recessive interaction is operative.

POLYGENIC INHERITANCE

Mendelism deals with classical genetics and is restricted to one or two genes, which control a single character. Later, it was discovered that certain characters were contributed collectively by several genes. This type of control was called as multiple factor inheritance or polygenic inheritance. Statistical analysis of a quantitative character can help in the study of polygenes.

If there are several genes influencing a single character, the character may show continuous variation. Since every allele accounts for small differences in the trait, all the alleles together contribute to the development of this quantitative trait. Examples of such characters are: Kernel colour in wheat, skin colour in man, ear size in corn, corolla length in tobacco etc.

Several methods of statistical analysis can be applied to study the quantitative traits. These are explained in later pages under the section: Normal distribution.

SEX RELATED INHERITANCE

Animals are separated in terms of sex as male and female. However, most plants have both male and female portions in the same flower though in monoecious plants, the male and female flowers are distinctly separate on the same plant (cucurbits) and dioecious plants like Asparagus, date palm etc have male and female elements in different individuals.

Sex provides for the great amount of genetic variability in natural populations. Several types of sex determining mechanisms have been classified in different organisms.

Sex chromosome mechanisms

Heterogametic males

Found in humans and almost all mammals. Males are determined by the Y chromosome (XY) Males produce two kinds of gametes X and Y. Hence they are called heterogametic. Some insects do not have Y chromosome in males (XO) and are heterogametic.

Heterogametic females

Found in most insects, butterflies, moths, silkworms, some birds and fishes. Single X chromosome (XO) determines females and two X chromosomes (XX) determine males. In some chicken species, there is a Y chromosome in females (XY). In the other species there is thus an XO female mechanism.

Genic balance

In Drosophila, the Y chromosome is not concerned with determination of sex, but the factors for maleness present in all of the autosomes collectively out number the femaleness factors present on X chromosomes. Hence one set of autosomes will have a value of male determining factors of 1 and one X chromosome has a value of female determining factors of 1 1/2. A normal male has AA XY and the ratio of male : female determinants is 2 : 1 1/2 hence it is a male. A normal female AA XX has a male : female determinants ratio of 2 : 3 hence it is a female. Several abnormal combinations also occur.

Haplodiploidy

Observed in bees. The male bees develop from unfertilized eggs and are haploid. Females (workers and queens) develop from fertilized (diploid) eggs. Sex chromosomes are not involved in sex determination in them (most Hymenoptera). The quantity and quality of food consumed by the diploid organisms will determine, if workers or queens should develop. Workers are sterile (environmental influence). Most of the eggs that the queen lays and fertilizes (from the stored sperm receptacle) will develop into workers and whichever eggs she chooses not to fertilize will develop into haploid males.

Single gene effects

Several other sex determining mechanisms are known e.g complementary sex factors, the transformer gene of Drosophila and mating type in micro organisms.

Inheritance related to sex

Several types of sex related inheritance patterns have been studied. They are:

Sex linked inheritance

Genes located on the X-chromosome in mammals, Drosophila etc are said to be sex linked (or X-linked). The genes located on the non-homologous portion of the Y- chromosome in humans are called Y-linked or holandric genes. These genes get transmitted from father to son only.

Sex-influenced traits

The genes governing sex-influenced traits may be located on auto somes or homologous portions of the sex chromosomes. These are mostly influenced by the internal environment created by the sex hormones. These are mostly found in higher animals and humans e.g. pattern baldness in humans. The gene exhibits dominance in men and recessiveness in women.

Sex-limited traits

Some genes express in one of the sexes due to hormonal environment or anatomical differences, e.g. genes for milk production are expressed only in the females. Chickens have a recessive gene for cock feathering in males but the same gene produces hen feathering in females.

Sexuality in plants

Most flowering plants are monoecious and have male and female flowers on the same plants but they do not have sex chromosomes. The fact that mitotically produced similar cells produce different sexual functions (male or female flowers or anthers and ovaries) proves that this is under the genetic control of a single gene locus. However, sex chromosomes have been identified in *Melandrium* where the Y chromosome determines maleness similar to humans. Self-fertilization is also a common feature in plants except for a few moneocious sps with a built in self-incompatibility to favour cross-fertilization.

CHROMOSOME MAPPING

Genes on different chromosomes are distributed into gametes independently of one another but genes on the same chromosome tend to stay together during the formation of gametes. Such genes located on the same chromosome are said to be linked. These genes stay together but get separated during meiotic crossing over since they get exchanged at chaismata between non-sister chromatids of the homologous pair of chromosomes.

If the two dominant alleles are on one chromosome and two recessive alleles on the other (AB/ab), it is called coupling phase. If the dominant allele of one locus and the recessive allele of the other occupy the same chromosome (Ab/aB), the relationship is termed repulsion phase. With such differences in linkage, the parental and recombinant gametes will be of different types.

Coupling parent:		Repulsion parent	
AB/ab		Ab/aB	
Gametes: parental	= AB & ab	Parental gametes	= Ab & aB
Recombinant	= Ab & aB	Rrecombinant gametes	= AB & ab

The farther apart 2 genes are located on a chromosome, the greater the opportunity for a chiasma to occur between them. When two strand double cross overs occur between two genes, a third gene locus between them is used to detect the double cross overs.

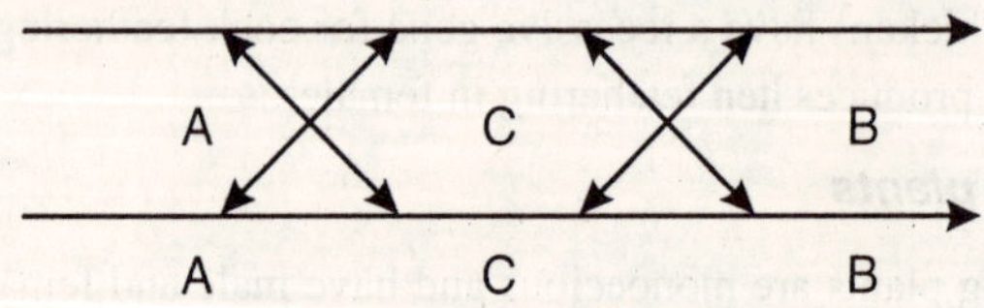

Genes reside in the chromosome in a linear order. To map these genes, the order of arrangement and relative distances between the genes should be found out. One map unit (centimorgan) is equivalent to 1% crossing over.

Procedure of finding the gene order and mapping

Test cross progeny is the best material to detect cross over gametes. Three point test crosses are preferred (which involve three genes and two cross overs plus a double cross over). Trihybrid individuals are generally test crossed to produce eight types of individuals in the progeny: Two parental types, two cross over region I types, two cross over region II types and two double cross over types.

P	Trihybrid parent	×	Test cross parent
	$\frac{Abc}{aBc}$		$\frac{abc}{abc}$

The gametes from the trihybrid are of 8 different types :

Abc, aBC, ABC, abc, AbC, aBc, Abc, abC

F_1	$\frac{Aa\,bb\,cc}{aa\,Bb\,Cc}$	$\frac{Aa\,Bb\,Cc}{aa\,bb\,cc}$	$\frac{Aa\,bb\,Cc}{aa\,Bb\,cc}$	$\frac{Aa\,Bb\,cc}{aa\,bb\,Cc}$
	Parentals (highest number)	Cross over region I	cross over region II	doble cross over (least numbers)

To find out the gene order, we need to screen the three possible orders depending upon the double cross over individuals (DCO) individuals. In the given populations of segregating individuals, the least pair of values will be the DCO. With this information, three trials can be carried out assuming one of the three genes in the middle. This is done by reading out the double cross over patterns. The correct one is selected as the combination matching with the least numbers (i.e. DCOs).

Interference and coincidence

The formation of one chaisma reduces the chances of another chiasma at the same time. This inability is expressed as fewer DCO types and is called **interference.** This is expressed in terms of coefficient of coincidence, which can be easily calculated.

$$\text{Coefficient of coincidence} = \frac{\%\text{ observed D.C.O}}{\%\text{ of expected D.C.O}}$$

Coincidence + interference = 1.0

i.e. Interference = 1 - Coincidence.

Problems

Problem 1

On one chromosome of tomato there are three recessive genes: **a** (causing absence of anthocyanin) **hl** (producing hairless plants) and **j** (producing jointless pedicels). The progeny from a trihybrid test cross contained the following phenotypes.

269 anthocyaninless, jointless	259 hairless
268 anthocyaninless,jointless, hairless	40 jointless, hairless
931 jointless	260 normal
941 anthocyaninless, hairless.	32 anthocyaninless

Find out the original linkage of the genes and estimate the map distances between them.

Solution

The most frequent phenotypes are parental types, i.e 941 anthocyaninless, hairless and 931 jointless. This means that **j** was on one homologous chromosome and **a** and **hl** were on the other. Now the gene order has to be found out.

The highest pair of numbers are those of the parental genotypes .The least numbers are D.C.Os (32 anthocyaninless and 40 jointless, hairless).

Trials for gene order

(1) $P = \dfrac{\text{AiHL}}{\text{aJhl}} \quad \text{DCO} = \dfrac{\text{AiHL}}{\text{aJhl}} = \dfrac{\text{Normal}}{\text{Jointless,hairless}}$

This order of the genes does not match the phenotypes of the DCOs.

(2) $P = \dfrac{\text{jhla}}{\text{jHLA}} \quad \text{DCO} = \dfrac{\text{JHLa}}{\text{jhlA}} = \dfrac{\text{anthocyaninless}}{\text{Jointless,hairless}}$

This order matches with the given DCOs phenotypes.

Therefore this is the correct gene order.

C.O --I (J HL A and j hl a)

260 + 268 + 32 + 40 (D.C.Os)

= 600 /3000 = 0.2 = 20% or 20 map units between j and hl.

C.O--II (J hl A and j HL a)

259 + 269 + 32 + 40 (D.C.Os)

= 600/3000 = 0.2 = 20% or 20 map units between hl and a.

Therefore the gene order and map distances are: j — 20 — hl — 20 — a

Problem 2

In Drosophila, a kidney bean shaped eye is produced by a recessive gene **k**, cardinal eye colour is produced by a recessive gene **cd.** Both are located on the same chromosome and in between them a third gene **c** producing ebony body colour is located. Homozygous kidney cardinal females are mated to homozygous ebony males. The trihybrid is test crossed to produce the F_2. Determine the map distances between the genes, from the following progeny:

128 kidney, ebony	1773 ebony
138 cardinal	1761 kidney, cardinal 8 wild type
97 kidney	89 ebony cardinal
6 kidney ebony cardinal	6 kidney ebony cardinal

Solution

The map distances between the genes are: K 7 e 5 cd

Problem 3

In corn, a single chromosome has three genes located on it. A recessive allele **c** produces colourless aleurone; the recessive **sh** produces shrunken kernels and a dominant gene **Wx** produces normal starchy endosperm while the recessive gene **wx** produces waxy endosperm. A trihybrid F_1 was produced by crossing a homozygous plant from a seed with colourless, plump and waxy endosperm and a plant from a seed with coloured, shrunken and starchy endosperm. The trihybrid was test crossed to a colourless, shrunken, waxy strain. The progeny exhibiting eight different phenotypes is given below. Find out the map distances between the genes and calculate the interference in the region.

he gene order is **c sh wx**

2 colourless, shrunken, waxy	113 clourless, shrunken, starchy
4 colourled, plump, starchy	116 colourled, plump, waxy
626 colourless, plump, starchy	2538 coloured, shruken, strarchy
601 coloured, shrunken, waxy.	2708 colourless, plump, waxy

Solution

The map distances are: c — 3.5 — sh — 18.3 — wx

The interference in the region is 86%.

MUTATION BREEDING

Mutations are defined as sudden heritable changes in an organism other than those due to Mendelian recombination. A change in the gene results in a mutation, which is manifested as an alteration of morphological or biochemical characteristics. Mutations can occur spontaneously in nature. However, they can also be induced artificially by mutagens. Physical mutagens like X-rays, gamma rays, ultraviolet rays etc and chemical mutagens like ethyl methane sulphonate, diethyl sulphate, etc can induce mutations in organisms. Several mutant varieties of crop plants have been produced through mutation breeding.

Methodology of mutation breeding

Seeds or seedlings (cuttings in case of vegetatively propagated plants) are exposed to either physical radiations or chemical mutagens by using doses / concentrations below LD 50 level. The LD 50 level for each mutagenic agent is determined by preliminary trials using several different doses / concentrations and recording the germination and survival rate. The dose / concentration where 50% of the seedlings / plants / cuttings die is determined as the LD 50. Doses / concentrations below the LD 50 are then chosen to initiate the mutation breeding process. The material is either irradiated (in case of the physical mutagens) or soaked in mutagenic solutions before being sown /grown. The first generation after the mutagenic treatment is known as the M1 generation. Seeds from normal looking plants from each treatment (each dose / concentration) along with the control are collected to raise the M2 generation. Screening the M2 progeny for mutants is carried out. These mutants are recessive. Further testing is needed to ascertain their true breeding nature. This is carried out by raising M4 and M5 generations if needed. The mutants can then be submitted to the seed testing and certification authorities to be considered for release as a variety. However, if the mutants are superior for only one character, but have a poor yield, this character can be transferred to the high yielding varieties by crossing. A mutation breeding experiment can be designed for any crop by careful selection of mutagens and treatment protocols. The breeder should have clear objectives and needs to be alert to isolate desirable mutants. In case of cross-pollinating crops, special care is taken to breed them appropriately (Poehlman 1987).

STATISTICAL ANALYSIS

The binomial expansion

When two independent events are occurring with the probabilities **p** and q**,** then the probability of their joint occurrence is the product of the independent events i.e. **pq.**

In $(\mathbf{p} + \mathbf{q})^n$, the **p** and **q** represent the probabilities of alternative events (and the number of trials is n). The sum of the factors in the binomial adds to unity.

$$\therefore p + q = 1$$

e.g. There are four possibilities possible if a coin is tossed twice (**1/2 = heads; 1/2 tails**).

First toss	Second toss		Probability
Tails (q)	and Heads (p)	=	pq
Tails (q)	and Tails (q)	=	q^2
Heads (p)	and Heads (p)	=	p^2
Heads (p)	and Tails (q)	=	Pq
			1

$(p + q)^2 = 1$

$p^2 + 2pq + q^2 = 1$ *(Similarly $(p + q)^3$ is also worked out).*

If there are two possible outcomes of a random experiment, they are called success (s) and failure (f).

Let P(s) = p and

P (f) = q

p + q = 1

In **'n'** trials, the total number of possible ways of obtaining **'r'** success and failure **(n - r)** is:-

Probability of **'r'** successes out of **'n'** trials = $P(r) = \frac{n!}{r!(n-r)!} . p^r q^{n-r}$

(where ! = factorial).

Problems on Binomial distribution

Problem 1

Determine the probability of obtaining 2 heads and 2 tails in 4 tosses of a coin.

Solution

Head = 2

Tail = 2

N = 4

Probability of 2 successes and 2 failures in 4 trails = $\frac{n!}{r!(n-r)!} . p^r q^{n-r}$

If the probability of obtaining a head (p) is equal to the probability of obtaining a tail, ***(then p = 1/2 and q = 1/2)***

$$= \frac{24}{4} \times \left(\frac{1}{2}\right)^4$$

$$= \frac{24}{4} \times \frac{1}{16} = \frac{24}{64} = 0.375$$

The probability of obtaining 2 heads and 2 tails in 4 tosses is 0.375.

Problem 2

What are the chances of having three boys and two girls in a family? The probability for the birth of a girl is **1/2** and the probability for the birth of a boy is **1/2.**

Probability of having 3 boys in the family $P(r) = \frac{n!}{r!(n-r)!} . p^r q^{n-r}$

$$P(3) = \frac{5!}{3!(5-3)!} . \left(\frac{1}{2}\right)^3 . \left(\frac{1}{2}\right)^2$$

$$P(3) = \frac{5 \times 4 \times 3 \times 2 \times 1}{3 \times 2 \times 1 \times 2 \times 1} . \left(\frac{1}{2}\right)^5$$

$$P(3) = \frac{120}{12} . \frac{1}{32} = \frac{10}{32}$$

$$P(3) = 0.312$$

Probability of having 3 boys in the family = **0.312**

Probability of having 2 girls in the family = 1 - 0.312 = **0.688**

It can also be represented in percentage **Boys = 31% and Girls = 69%.**

The Poisson distribution

This studies the probability of rare events like a rare disease, mutations etc. Here the **p** is very small in a large **n** i.e the probability of success for every trial is very small though the **n** is very large. Analysis of data by Poisson distribution (i.e. to find out the probability of events occurring) is suitable in cases where the **p** of a single event nor **n** is known since it makes use of **np.** **np** = μ which is constant and is the mean of the distribution (the values of **e**-μ where **e** = 2.7183, a constant, are read from the table).

$$\text{Probablity of r successes} = \frac{e_{-\mu} . \mu_r}{r!}$$

$$P(r) = \frac{e^{-\mu} \cdot \mu^{\rho}}{r!}$$

where μ = mean value given in the problem (np).
$e^{-\mu}$ = value read from the table under μ
r = successes 0,1,2,3, etc.

μ	01	1	2	3	4	5	6	7	8	9	10
e-μ	0.9048	0.3679	0.1353	0.0498	0.0183	0.0025	0.0025	0.0009	0.0003	0.0001	0.000045

Problems on Poisson distribution

Problem 1

In a population study of 100 individuals, calculate by Poisson distribution the probabilities that it contains (a) 1 albino (b) 2 albinos.

Suppose that 1 out of 1000 individuals is an albino.

Solution

n = sample = 100

Probability of albinos = 1 in 1000 i.e $\frac{1}{1000} = 0.001$

Average = μ = probable number of albinos

= np 100 × 0.001 = 0.1

r = events 0, 1, 2 albinos.

(a) r = 1

Substitute in p(r) $\frac{e^{-\mu} \cdot \mu^{r}}{r!}$

Where μ = mean value = 0.1
$e^{-\mu}$ = from table under 0.1 = 0.9048
r = 1

$$p(r) = \frac{0.9048 \text{ x } (0.1)^1}{1!} = \frac{0.9048 \text{ x } 0.1}{1}$$

0.09048 (probability of 1 albino).

(b) $r = 2$

$$p(r) = \frac{0.9\text{-}48 \times (0.1)^2}{2!} = \frac{0.9048 \times 0.01}{2 \times 1} =$$

0.004524 (probability of 2 albinos).

Problem 2

An average of 5 cars pass under a bridge every minute. Calculate by Poisson distribution the probabilities that exactly 0, 1, 2, 3 and 4 cars pass in a one minute period.

Solution

$\mu = 5$

$e^{-\mu} = 0.0067$

$r = 0, 1, 2, 3$ and 4

$$r = 0: \; p(0) = \frac{0.0067 \times (5)^0}{0!}$$

$$= \frac{0.0067 \times 1}{1} = 0.0067$$ (the probability that no cars pass in a one minute period).

$$r = 1: p(1) = \frac{0.0067 \times (5)^1}{1!}$$

$$= \frac{0.0067 \times 5}{1}$$ 0.0335 (the probability that 1 car passes in a one minute period).

$$r = 2: p(2) = \frac{0.0067 \times (5)^2}{2!}$$

$$= \frac{0.0067 \times 25}{2 \times 1} = 0.08375$$ (the probability that 2 cars pass in a one minute period).

$$r = 3: p(3) = \frac{0.0067 \times (5)^3}{3!}$$

$$= \frac{0.0067 \times 125}{3 \times 2 \times 1} = 0.1395$$ (the probability that 3 cars pass in a one minute period).

$$r = 4: p(4) = \frac{0.0067 \times (5)^4}{4!}$$

$$= \frac{0.0067 \times 625}{4 \times 3 \times 2 \times 1} = 0.1744$$ (the probability that 4 cars pass in a one minute period).

The Normal distribution

The study of quantitative traits in a large population usually reveals that ver few individuals posess the extreme phenotypes and that progressively mor number of individuals are found nearer the average value for the populatior This reveals a bell shaped curve called **Normal distribution.**

Since it is very difficult to measure each indvidual in a population (e.g 100 individuals) a random sample of a few numbers is drawn and calculatior done on this sample. The sample is treated as a true representative of th large sample and its mean (average) is referred as **X** which should be equ to μ (mean of the actual population).

$$\text{Mean} = X = \frac{\Sigma x}{n}$$

Where Σ means added value,

X means the variable, that is being measured,

N the number of individuals in the sample.

Quantitative traits include the economically important traits such as bo weight gains, mature plant heights, egg or milk production records, yield grain per acre, seeds per pod or seed weight. These are quantitative or met traits with continuous variability (i.e in a population, each individual has different estimate). These traits may be governed by polygenes and mo than one gene will contribute to a given trait. The mean value gives t average but does not give an estimate of the variability. Populations with t same mean values differ in their degree of variability. Hence one of the m useful measures of variability in a population for genetic purposes is the **standard deviation**, represented by sigma (σ).

The standard deviation of the sample is represented by σ.

$$\sigma = \sqrt{\frac{\Sigma(x - X)^2}{n - 1}}$$

The square of standard deviation is called **variance** (σ^2). The standard deviation can be expressed as a percentage of the mean by the **coefficient of variation.**

$$\text{Coefficient of variation} = \frac{\text{Standard deviation}}{X} \times 100$$

$$\text{or C.V} = \frac{\sigma}{X} \times 100$$

Standard error is a quantity which can be calculated directly from standard deviation of the sample and sample size.

$$\text{Standard error of the mean(SEM)} = \frac{\text{Standard deviation}}{\sqrt{n}} = \frac{\sigma}{\sqrt{n}}$$

Where n = sample population

SEM can be related to the population. The value decreases with increase in the number of observations. The sample mean is hence related to the population mean with the help of confidence intervals. For this, the degrees of freedom (**n-1**) has to be determined and the value of **t** found out from the table against d.f and probability level (of **p = 0.05** meaning that **5%** chance of being wrong and 95% chance of being right).

Multiplying this **t** value with the **SEM**, the confidence limits are obtained i.e the quantity that deviates on both sides of the sample mean.

t (0.05p) x SEM value = Confidence limits

e.g. Mean = X = 7

Standard deviation = 1.25

$$SEM = \frac{1.25}{\sqrt{10}} = \frac{1.26}{3.16} = 0.39$$

d.f (n-1) = 10 – 1 = 9.

t value from table at 0.05 p for 9 d.f = 2.26.

Confidence limits to the mean = t x SEM = 2.26 x 0.39 = 0.88

Therefore the mean can be written as 7 ± 0.88.

Therefore, the mean value ranges from 6.12 to 7.88 at 0.05 p level.

Problems on Normal distribution

Problem 1

Calculate the mean, standard deviation, variance and standard error of the mean for the following data:

variable (x) : 10, 13, 17, 22, 27, 30, 31, 32

x	(x – X)	(x – X)2
10	- 12.7	161.29
13	- 9.7	94.09
17	- 5.7	32.49
22	- 0.7	0.49
27	4.3	18.49
30	7.3	53.29
31	8.3	68.89
32	9.3	86.49
Σx = 182	**n = 8**	**$\Sigma(x - X)^2$ = 515.52**

$$\text{Mean} = X = \frac{\Sigma x.}{n} = \frac{182.}{8} = 22.7$$

$$\text{Variance} = \sigma^2 = \frac{\Sigma\ (x - X)^2}{n - 1} = \frac{515.52}{8 - 1} = 73.65$$

$$\text{Standard deviation} = \sigma = \sqrt{73.65} = 8.58$$

$$\text{Standard error} = \frac{\sigma.}{\sqrt{n}} = \frac{8.58\,.}{\sqrt{8}} = \frac{8.58\,.}{2.82} = 3.04$$

Therefore the mean can be written more accurately as 22.7 ± 3.04

Mean = 22.7

Variance = 73.65

Standard deviation = 8. 58.

Problem 2

Compute the values of coefficient of variation from the following data, and compare the two series for the data consistency.

Series A : X = 34.00 σ = 12.00

Series B : X = 27.00 σ = 6.00

$$C.V = \frac{\sigma}{X} \text{ X } 100$$

$$C.V\ (\text{Series A}) = \frac{12.00}{34.00} \text{ x } 100 = 35.20\%$$

$$C.V\ (\text{Series B}) = \frac{6.00}{27.00} \text{ x } 100 = 22.22\%$$

This shows that the data of Series B are more variable and less Sconsistent as compared to the data of Series 'A' which are more consistent.

The consistency is deduced from the fact that the mean value is close to the C.V value.

Comparison of means of two samples

Comparison of two means can carried out by applying tests of significance. This helps in reaching a conclusion whether the difference is significant or whether it is not significant hence may have occured due to chance. At first a hypothesis is laid: There is no difference between the sample means, i.e hypothesis of no difference or null hypothesis.

H O μ = X (no difference)

μ = Population mean

X = Sample mean.

Rejection of hypothesis means that there is a difference between the two means. Acceptance of hypothesis means there is **no** difference. The level of significance chosen is generally 0.05p.

There are several tests available (the t test and the Chi-Square test):

The 't' - test

This test put forward by an author who called himself **Student** is also called the student's **'t'** test. This is used to compare two samples.

E.g. To find out whether there is a significant difference in plant height due to treatment with a particular fertilizer , the data from the control and treated **populations is collected and mean values calculated.**

$$X \pm S.E$$

$$\text{Control} = 68 \pm 2.03 \quad n = 10$$

$$\text{Treated} = 96 \pm 0.06 \quad n = 9$$

$$t = \frac{\text{Mean Difference}}{\text{Summation of square of standard errors of control and treated}} = \frac{X^c - X^t}{(S.E^c)^2 + (S.E^t)^2}$$

at $n^c + n^t - 2$ degrees of freedom (df).

$$t = \frac{68 - 96}{(2.03)^2 + (0.06)^2} \text{ at } 10 + 9 - 2 \text{ d.f}$$

$$t = \frac{-28}{4.1236} = 6.79 \text{ at } 17 \text{ d.f}$$

Calculated t value = 6.79 at 17 d.f

Tabulated 't' value = 2.11 at 17 d.f at 0.05p

(The 't' distribution table is consulted.)

The tabulated **'t'** value is lesser than the calculated **'t'** value. Therefore the null hypothesis is rejected and the difference between the two means is significant (the fertilizer had a significant effect). If the tabulated **'t'** value is greater than the calculated **'t'** value, the null hypothesis will be accepted and the difference between the two means will not be significant.

Problems on 't' test

Problem 1

In the data given below, find whether the two means are significantly different.

	X_1	X_2
n	30	35
Mean	32	38
Variance	9.62	14.23

Solution

$$t = \frac{X1 - X2}{(S.E_1)^2 + (S.E_2)^2} \text{ at } n_1 + n_2 - 2 \text{ d.f}$$

Variance (1) = 9.62	Variance (2) = 14.23
S.D(1) = 3.10	S.D(2) = 3.77

$$S.E\,(1) = \frac{\sigma}{\sqrt{n}} = \frac{3.10}{\sqrt{30}} = \frac{3.10}{5.47} \qquad S.E\,(2) = \frac{3.77}{\sqrt{35}} = \frac{3.77}{5.91}$$

$$t = \frac{32 - 38}{(0.56)^2 + (0.63)^2} \text{ at } 30 + 35 - 2$$

$$= \frac{-6}{0.3136 + 0.3969} \text{ at 63 d.f}$$

$$= \frac{-6}{0.7105} = 8.45 \text{ at 63 d.f}$$

The tabulated t value = 2.00 at 0.05 d.f

Since the tabulated 't' value is lesser than the calculated 't' value, the null hypothesis is rejected and the difference between the two means is significant.

Problem 2

In the data given below, find whether the two means are significantly different:

	x_1	x_2
Mean X	53.49	75.8
Standard deviation σ	7.037	5.385
N	51	30

$$\text{Standard error} = \frac{\sigma}{\sqrt{n}}$$

$$S.E_1 = \frac{7.037}{\sqrt{51}} \qquad S.E_2 = \frac{5.385}{\sqrt{30}}$$

$$= \frac{7.037}{7.14} = 0.98 \qquad = \frac{5.385}{5.47} = 0.98$$

$$t = \frac{X_1 - X_2}{(S.E_1)^2 + (S.E_2)^2} \text{ at n2 + n2 - 2 d.f}$$

$$= \frac{53.49 - 75.8}{(0.98)^2 + (0.98)^2} \text{ at } 51 + 30 - 2 = 79 \text{ d.f}$$

$$= \frac{22.31}{0.9604 + 0.9604} = \frac{22.31}{1.9208} = 11.614$$

Calculated 't' = 11.614

Tabulated 't' at 79 d.f = 3.37 (the appendix).

Thus, the calculated **'t'** far exceeds the tabulated **t** at 79 d.f and the null hypothesis that there is no difference between the two samples is rejected and the difference between the two samples is highly significant.

X^2 (Chi - square) test

This test used to find out whether there is any significant difference between the expected genetic ratios (e.g **3 : 1.**, **9 : 3 : 3 : 1** etc) and the observed ratios. This was explained in the earlier pages of this chapter also. The Chi-

Square test can be applied to problems concerning dihybrid ratios and also in the study of Hardy-Weinberg's equillibrium.

In the X^2 test the degrees of freedom are calculated from the number of classes involved. For e.g, in a mono hybrid cross, there are only two phenotypes (except in the incomplete dominance where there are three classes) and d.f = 2 – 1 = 1. Hence the Chi-Square test is used to evaluate a genetic hypothesis whether the observed deviations occur due to chance.

$$\text{The formula is : } X^2 = \frac{\Sigma (O - E)^2}{E}$$

where O represents the observed frequency and

E represents the expected frequency.

Problems on Chi-square test

Problem 1

In crosses of F2 generation, there were 785 tall and 270 short plants. Does this conform to the expected Mendel's ratio of 3 : 1?

Chi - Square $X^2 = \Sigma (D^2 / E)$

Where Σ = stands for sum of

D = deviation from expected ratio

= O – E

where O = observed value

E = expected value

Total progeny = 1055		
Hypothesis	= 3 tall	: 1 short
observed(o)	= 785	270
Expected(E)	= 789	263
Deviation [O - E] = (D)	= -4	7
D2	= 16	49
D2/E	= 16/789	49/263
	= 0.02	0.18

X2 = 0.02 + 0.18 = 0.20

Degrees of freedom (df) should now be calculated before the interpretation of the data to the 3 : 1 hypothesis. (df) is always 1 less than the number of classes. Here there are two classes (tall and short). Therefore n – 1 = 2 – 1 = 1 df. The Chi-Square table has to be screened to determine the probability (p) that our data fit the ratio and that the deviations observed can be attributed to chance.

Compare the calculated X^2 value with those given horizontally in the table for one degree of freedom. The tabulated X^2 value at **1 d.f** at **0.05 p** is **3.84.** Since the calculated X^2 value is lesser than the tabulated value, the null hypothesis is accepted and the difference between the observed and expected values is not significant.

Problem 2

A red flowered variety of 4 'O' clock plant was crossed with a white flowered variety, the F_2 progeny included 22 red; 52 pink; 23 white flowered plants. Ascertain whether they follow the Mendelian pattern of 1 : 2 : 1.

Solution

	Red	Pink	White	Total
O	22	52	23	97
E	24.25	48.50	24.25	
O-E	- 2.25	3.50	-1.25	
$(O-E)^2$	5.06	12.25	1.56	
$\frac{(O-E)^2}{E}$	$\frac{5.06}{24.25}$	$\frac{12.25}{24.25}$	$\frac{1.56}{24.25}$	
	0.21 +	0.25 +	0.06	= 0.53

X^2 = 0.53 at 3 – 1 = 2 d.f

The tabulated X^2 value at 2 d.f at 5% level of probability (0.05) is 5.99 (refer the X^2 table in the appendix). The calculated X^2 value is lesser than the tabulated value hence, the difference between the observed and expected values is not significant. The ratio follows the Mendelian pattern.

Problem 3

A dihybrid cross of pea plants resulted in the following progeny:-

Round yellow :	Round green :	Wrinkled yellow :	Wrinkled green
317	109	102	32

Calculate X^2 and draw conclusions.

Solution

	Round yellow	Round green	Wrinkled yellow	Wrinkle d green	Total
O	317	109	102	32	560
E	315	105	105	22.5	560
O – E	2	4	3	9.5	
(O – E)	4	16	9	90.25	
$\frac{(O-E)^2}{E}$	$\frac{4}{315}=0.012$	$\frac{16}{105}=0.152$	$\frac{9}{105}=0.085$	$\frac{90}{22.5}=4$	
	0.012		0.152	0.08	
$0.012 + 0.152 + 0.085 + 4 = 4.249$					
$X^2 = 4.249$ at $4 - 1 = 3$ d.f					

The tabulated X^2 at 3 d.f at 0.05 p is **7.82.** Since the calculated X^2 value is lesser than the tabulated value the null hypothesis is accepted and the difference between the observed and expected values is **not significant**. It indicates the independent assortment of genes.

Problem 4

In peas, the yellow cotyledon colour is dominant to green and inflated pod shape is dominant to constricted form of pod. The F_1 dihybrid was self fertilized to obtain the progeny : 193 green inflated : 184 yellow constricted : 556 yellow inflated : 61 green constricted. Calculate the X^2 and draw conclusions.

Solution

	yellow inflated	yellow constricted	green Inflated	green constricted	total
O	556	184	193	61	994
E	559.1	186.4	186.4	62.1	994
O-E	- 3.1	- 2.4	6.6	- 1.1	
(O-E)2	9.61	5.76	43.56	1.21	
$\frac{(O-E)^2}{0.019\ E}$	0.017	0.031		0.234	

0.017 + 0.031 + 0.234 +0.019 = 0.301.

X^2 = 0.301 at 4 - 1 = 3 d.f

Tabulated X^2 at 3 d.f at 0.05p is 7.82. Since the calculated X^2 is less than the tabulated value, the null hypothesis is accepted and the difference between the observed and expected values is not significant, indicating independent assotment of the genes according to the 9 : 3 : 3 : 1 ratio.

Correlation

The correlation coefficient (r) measures how closely two sets of data are associated has the limits of + 1 and has no units.e.g.:

(1) The heights of boys in a school and their respective weight

(2) Concentration of pesticides and their effect on germination rate.

$$\text{Correlation coeffient(r)} = \frac{\Sigma\ (x - X).(y - Y)}{\Sigma\ (x - X)^2.\Sigma\ (y - Y)^2}$$

Where x = independent variable represented on x axis.

Y = dependent variable on y axis.

(x – X) and (y – X) are deviations from the respective means.

N = represents the number of observations.

The corrrelation coefficient (r) varies between **–1 to +1.** A positive correlation is obtained when **r = +1** and a negative correlation when **r = –1.** When there is no correlation, **r = 0.** The degrees of freedom (df) for correlation coefficient are the number of pairs (x and y values) minus one pair. The significance of the correlation can be tested by looking up the **tabulated r** value in the r table at 0.001 at the particular d.f. To find out the level of significance at the population level, for more accuracy, a t - test may be applied to evaluate the ratio of **r** to the standard error of **r** with **n - 2 d.f.**

$$\text{S.E of r} = \frac{(1-r^2)}{n-2}.$$

$$\text{Thus } t = \frac{r}{(\text{S.E of r})} = \frac{r}{1-r^2 \;/\; n-2}$$

$$\text{i e } t = \frac{r\,(n-2).}{(1-r^2)} \quad \text{for } n-2 \text{ d.f}$$

The t value can be compared with the tabulated t and conclusions made.

Problems on correlation

Problem 1

In a study involving salinity tolerance of resistant cultivars of rice plants, the following results were obtained with regard to percentage of susceptible plants.

Sodium chloride (%) =	0.5	1	2	3	4	5	7	10
Number of susceptible plants =	1	2	5	8	11	16	19	28

Solution

x	y	x-X	$(x-X)^2$	y-Y	$(y-Y)^2$	(x-X).(y-Y)
0.5	1	3.56	12.67	0.25	105.06	36.49
1.0	2	3.06	9.36	9.25	85.56	28.30
2.0	5	2.06	4.24	6.25	39.06	12.87
3.0	8	1.06	1.12	3.25	10.56	3.445
4.0	11	0.06	0.003	0.25	0.0625	0.015
5.0	16	0.94	0.88	4.75	22.56	4.465
7.0	19	2.94	8.64	7.75	60.06	22.785
10.0	28	5.94	35.28	16.75	280.56	99.49
$\Sigma x = 32$	$\Sigma y = 90$		$\Sigma(x-X)^2 = 72.193$		$\Sigma(y-Y)^2 = 603.48$	$\Sigma(x-X)(y-Y) = 207.86$

$$X = \frac{32.5}{8} = 4.06 \qquad Y = \frac{90}{8} = 11.25$$

Using the formula : Correlation coefficient $r = \frac{\Sigma(x-X).(y-Y)}{\Sigma(x-X)^2 . (y-Y)^2}$

$$= \frac{207.86}{72.2 \times 603.5}$$

$$= \frac{207.86}{208.74} = 0.995 \text{ d.f} = 8 - 1 = 7$$

Tabulated r = 0.898 at 0.001 p

Since the calculated **r** value exceeds the tabulated value at 0.001 p at 7 d.f the correlation between the concentration of sodium chloride used and the number of plants susceptible to salinity is highly significant. For confirmation of the level of significance shown by the **r** value, the **t-test** is done :

$$t = \frac{r}{\text{S.E of r}} \text{ for n - 2 d.f}$$

$$\text{S.E of r} = (1 - r^2)/n - 2.$$

$$= \frac{1-(0.99)^2}{8 - 2} = \frac{(1-0.98)}{8 - 2}$$

$$= \frac{0.02}{6} = 0.0033 = 0.057$$

$$t = \frac{0.99.}{0.057} = 17.37 \text{ t value. (calculated t - value)}$$

The tabulated t - value (appendix). at 8 – 2 = 6 d.f at 0.001 p is 5.2

Since the calculated t - value exceeds the tabulated t - value, the level of significance exhibited by r value is confirmed by the t - test. i.e samp correlation agrees well with that of the population.

Problem 2

In an experiment relating to height of plants and number of primary branches, the following data was obtained:

S.No	1	2	3	4	5	6	7	8	9	10
Height (cm)	15	28	34	23	37	30	24	29	27	20
No of leaves	12	26	29	16	30	25	20	18	19	16

Compute the correlation coefficient and interpret it.

Solution

The correlation coefficient for 9 d.f. is 0.910.

Tabulated **(r)** for 9 d.f = 0.847 at 0.001p. Since the calculated **'r'** exceeds the tabulated **(r)** at 0.001p, the correlation between the two attributes **is highly significant**. The t value also can be calculated for further testing as shown in the previous problem.

POPULATION GENETICS (HARDY-WEINBERG'S EQUILIBRIUM)

Gene and genotype frequencies

A population that is sexually reproducing with a relatively close degree of genetic relationship (species) residing within defined geographic boundaries wherein inter breeding occurs is called a **Mendelian population**. A pool consisting of all gametes produced by a Mendelian population from which the next generation will arise is called the **gene pool.** However, the percentage of gametes in the gene pool bearing e.g. **(A)** or **(a)** will depend upon the **genotypic frequencies** of the parental generation whose gametes form the pool. The gene frequency therefore depends on the genotype frequency i.e if more number of recessive genotypes **(aa)** are present in the population, the frequency of **(a)** will be higher than **(A).** If the relative frequencies of **(A)** and **(a)** gametes in the gene pool are known, we can calculate (provided there is a random mating between the gametes) the expected frequencies of progeny genotypes and phenotypes.

If p = percentage of (A) alleles in the gene pool and

q = percentage of (a) alleles,

then we can deduce all the possible chance combinations of these gametes.

	p (A)	q (a)
P (A)	p^2 AA	pq aA
Q (a)	pq Aa	q^2 aa

p + q = 1

The expected genotypic (zygotic) frequencies in the next generation may be summarized as follows:

$$(p + q)_2 = p_2 + 2pq + q_2$$

AA Aa aa

This comprises the whole of the population and hence all of these genotypic fractions must add to unity. This formula expressing the genotypic expectations of progeny in terms of gametic (allelic) frequencies of the parental gene pool is called the **Hardy -Weinberg Law:**

The Hardy-Weinberg Law states that a population that is infinitely large and mates at random (panmictic) and where there is no selection, migration or mutation, is said to be in **equilibrium.**

Calculation of gene frequencies

The method of calculating gene frequencies differs with different types of loci.

(a) Auto somal loci with two alleles

Codominant autosomal alleles

example: A_1A_1 = homozygous for one allele (D)

A_1A_2 = heterozygous (H)

A_2A_2 = homozygous for other allele (R)

Therefore N = D + H + R

Since they are diploid, the total number of individuals will have 2N alleles among which A1A1 genotype has two A1 alleles (2D). Heterozygotes (H) have only one A1 allele and one A2 allele. A2A2 genotype has two A2 alleles (2R).

If p = frequency of A1 allele and

q = frequency of A2 allele.

$$p = \frac{2D + H.}{N} = \frac{D + 1/2\,H}{2N}$$

$$q = \frac{H + 2R.}{N} = \frac{1/2H + R.}{2N}$$

Dominant and recessive autosomal alleles

In this case, the method of calculation is different from that of co-dominant alleles. We cannot distinguish between (**AA**) and (**Aa**) since both have the same phenotype. However we can distinguish the (**aa**) phenotype with certainty. In such a situation, we can get the estimate of **q** (the frequency of the recessive allele) from $\mathbf{q}^2$ (the frequency of the recessive genotype).

Sex influenced traits

The expression of dominance and recessiveness of genes will change due to environmental influence like sex hormonal effect. This happens in sex influenced traits. The homozygous genotypes will behave in a similar way in both sexer but the heterozygous genotype will produce different phenotypes in the two sexes in reverse order. The allelic frequency of the recessive phenotype is determined indirectly ($q = q^2$). A similar approach in the opposite sex gives an estimate of p. However p + q should be 1.

(b) Auto somal loci with multiple alleles

If **A>aı >a** is the dominance heirarchy in multiple alleles and their respective frequencies **p,q** and **r,** then the random mating will result in the following zygotic frequencies:

$$(p + q + r)^2 = p^2 + 2pq + 2pr + q^2 + 2qr + r^2 = 1$$

$$\underbrace{\frac{AA\ \ Aa_1\ \ Aa.}{A}} \qquad \frac{a_1a_1\ \ a_1a}{a_1} \qquad \frac{aa}{a}$$

(c) Sex linked loci

In these loci the method of calculation of allelic frequencies differs in co-dominant and dominant recessive sex linked alleles.

Co-dominant sex-linked alleles

Data from both males and females can be used but in organisms with **XY** mechanisms, the heterozygous condition is observed only in females while males are hemizygous.

Dominant and recessive sex-linked alleles

Since there is only one sex-linked allele in males, the frequency of a sex-linked trait among males is a direct measure of the allelic frequency in the population.

(d) Testing a locus for equilibrium

We cannot distinguish between the homozygous dominant and heterozygous class hence only co-dominant alleles can be tested for equilibrium through X^2 test.

Degrees of freedom

The degrees of freedom, **df** = **n - 1**. It represents the number of phenotypes minus the number of alleles.

Problems in population genetics

Problem 1

If the frequency of gene **A** is 0.3 and that of gene **B** is 0.5, find the equilibrium frequencies of the gametes **AB, Ab, aB and ab.**

Solution

For the gene A, p = 0.3 and q = 0.7 (a)

For the gene B, p = 0.5 and q = 0.5 (b)

Hence the equilibrium frequencies of the following are:

AB = (0.3) x (0.5) = 0.15

Ab = (0.3) x (0.5) = 0.15

AB = (0.7) x (0.5) = 0.35

Ab = (0.7) x (0.5) = 0.35

Problem 2

A legume population is segregating for the colours golden, light green and dark green produced by the codominant genotypes **$C_G C_G$, $C_G C_D$, $C_D C_D$** respectively. A sample population contained 2 golden, 36 light green and 162 dark green. What are the frequencies of the alleles C_G and C_D ?

Solution

Assume the frequencies of C_G and C_D as p and q respectively.

$$p = \frac{D + 1/2H}{N} = \frac{2 + 1/2(36)}{200} = \frac{2 + 18}{200} = \frac{20}{200} = \mathbf{0.1}$$

$$q = \frac{1/2H + R}{N} = \frac{1/2(36) + 162}{200} = \frac{18 + 162}{200} = \frac{180}{200} = \mathbf{0.9}$$

Frequency of C_G = 0.1

Frequency of C_D = 0.9

Alternatively: Frequency of q = 1 - p = 1 - 0.1 = 0.9.

Problem 3

Out of a total of 900 sheep of a particular breed maintained on a farm, 891 sheep were white and 9 black. Estimate the allelic frequencies. (White wool is produced by the dominant allele B and black wool by its recessive allele b).

Solution

We assume that the population would be in equillibrium

$$p^2 (BB) + 2pq (Bb) + q^2 (bb) = 1.0$$

The square root of the recessive phenotype would give an estimate of q.

$$q = q^2 = \frac{9}{900} = 0.1 = \text{frequency of allele b.}$$

$$p + q = 1 \quad \text{Therefore } p = 1 - q$$

frequency of allele A = 1 - 0.1 0.9.

Problem 4

Certain people have the ability to taste the chemical phenyl thiocarbamate (PTC). This taster ability is governed by a dominant allele **T**. The recessive allele **t** governs the inability to taste PTC. If 25% of a population are homozygous tasters and 44% are heterozygous tasters, what is the frequency of **t**? Let the homozygous tasters **TT** be represented as D (dominant). Let the heterozygous tasters **Tt** be H (heterozygous). Let the homozygous non-tasters **tt** be R (recessive). If **p** is the frequency of the allele **T** and **q** is the frequency of the allele **t,** then

$$p = \frac{D+1/2H}{N} = \frac{25+1/2(44)}{100} = \frac{25+22}{100} = \frac{47}{100} = \mathbf{0.47}$$

$$q = \frac{1/2H+R}{N} = \frac{1/2(44)+31}{100} = \frac{22+31}{100} = \frac{53}{100} = \mathbf{0.53}$$

The frequency of **t** is therefore **0.53.**

Alternative method :q = 1 - p (1 - 0.47 = 0.53).

Problem 5

Baldness is governed by a sex-influenced trait that is dominant in men and recessive in women. In a sample of 10,000 men, 7500 were found to be nonbald. In a sample of women of 10,000, how many non-bald women are expected ?

Solution

Genotype	Phenotype Males	Phenotype Females
$b_1\,b_1$	Bald	Bald
b^1b_2	Bald	Non-bald
$b_2\,b_2$	Non-bald	Non-bald

Let p = frequency of b_1 allele and

Q = frequency of b_2 allele.

$$\underset{(b_1\,b_1)}{p^2} + \underset{(b_1b_2)}{2\,pq} + \underset{(b_2\,b_2)}{q^2} = 1.0$$

In men, the allele b_2 is recessive (non-bald)

Then $q = q^2 = \frac{7500}{10,000} = 0.7500 = 0.75$

In women, the allele b_1 is recessive (bald)

Then $p^2 = (0.25)^2 = 0.0625$ or 6.25%

6.25% of the women of this population of 10,000 (i.e 625) will be bald.

Hence 9375 women out of 10,000 will be non-bald.

Problem 6

A sex-linked recessive gene controls colour blindness in humans. In a survey of a population of 1000 men, 40 were colour blind. (a) What is the gene frequency of the normal allele in the population ? (b) What percentage of the females in this population would be expected to be normal ? The gene is located on the X chromosome.

Solution

a)

Observed number of men	Genotypes of men	Phenotypes of men
960	C^+Y	Normal
40	c Y	Colour blind
1000		

Total number of X chromosomes = 1000.

40 X chromosomes carry the gene for colour blindness c

$q = 40/1000 = 0.04$ or 4% c alleles

$p = 1 - q = 1 - 0.04 = 0.96$ or 96% C^+ alleles

The gene frequency of the normal alleles in this population is 96% or 0.96.

(b) ***Since females posess two X chromosomes, the formula $(p + q)^2$ should be used.***

$$p^2 + 2pq + q^2 = 1.0 \text{ (100\% females)}$$

$$C^+C^+ \qquad C^+c \qquad c\,c$$

To find out the frequency of normal females, the frequency of colour blind females has to be found out.

Substitute $q^2 = (0.04)^2 = 0.0016 =$ **0.16% Colour blind females.**

The percentage of normal females = 100% - 0.16% = 99.84%.

The percentage of normal females in the population = 99.84%.

REFERENCES

1. Avers. C. (1982) Cell Biology. 2nd edition. D. Van. Nostrand. New York.
2. Avers. C. (1982) Genetics. 2nd edition. D. Van. Nostrand. New York.
3. Dobzhansky. T. F. J.. Ayala. F., Stebbins. G. L. and Valentine. J. W. (1977) Evolution. Freeman and Co. San Francisco.
4. Fisher. R. A. and Yates. F. (1963) Statistical tables for biological, agricultural and medical research. 6th edition. Oliver and Boyd. Edinburgh.
5. Griffiths. A. J. F., Miller. J. H., Suzuki. D. T., Lewontin and Gelbart. W. M (1993) An introduction to Genetic Analysis. 5th edition. Freeman and Company. New York.
6. Holleander. A. and de Serres. F. J. (1978) Eds. Chemical mutagens: Principles and methods for their detection. Vol. 5. Plenum Press. New York.
7. Poehlman. J. M. (1987) Breeding Field Crops. 3rd edition.Van Nostrand Company. New York.
8. Sokal. R. R. and Rohlf. F. J. (1969) Biometry. W. H. Freeman. San Francisco.
9. Stern. C. and Sherwood. E. R. (1966) The origin of Genetics. A Mendel source book: W. H. Freeman and Co. New York.

5

Plant Molecular Genetics

Tremendous improvements have been achieved in crop performance by adopting conventional and mutational breeding strategies. However, these strategies have their own limitations since traits present in unrelated species cannot be combined due to crossing incompatibility barriers. Further, the genotypes produced by induced mutations need extensive field testing before being released as new varieties. Even somatic hybridization has many limitations in terms of yielding viable hybrids. Generally, cross breeding combines all the genes of both parents thus resulting in a re-assortment of both desirable and undesirable traits. Several generations of back crossing will then be required to eliminate the undesirable traits which is time consuming and gets delayed results. Gene mapping by the use of conventional cross breeding methods in an attempt to assign genes to particular chromosomes and to clearly determine the distances between them have been explained in chapter 4. Some of the limitations of traditional plant breeding are overcome by using the techniques of molecular genetics which include gene cloning and restriction mapping. Analysis of restriction fragment length polymorphisms (RFLPs) allows mapping of genes even in cases where the phenotypic expression of the gene is unknown. Restriction endonucleases can be applied to a following number of molecular procedures:

(a) **Identification of the specific DNA sequence:** Each DNA sequence has a unique arrangement of restriction sites which define a physical map. These can be used as a fingerprint of DNA structure through restriction fragment length polymorphisms (Southern blotting).

(b) **Gene cloning:** The insertion and propagation of discrete segments of DNA into plasmid, cosmid, yeast artificial chromosomes (YAC) or bacterial artificial chromosomes (BAC) vectors depends upon restriction enzymes to create a specific cut site in the vector where the desired DNA will be inserted.

(c) **Gene sequencing:** DNA sequencing depends upon the use of restriction enzymes and involves the detection of the arrangement and sequence of nucleotide bases. Sequencing data can, of course, reveal the presence of additional restriction endonuclease recognition sites.

AN OVERVIEW OF THE METHODS OF MOLECULAR GENETICS

After its isolation, DNA is digested with suitable restriction endonuclease enzymes to yield distinct fragments. The DNA fragments can be separated on the basis of their size by gel electrophoresis. This results in a series of bands corresponding to fragments of particular size which progressively decrease down the gel. The size of these fragments can be callibrated against marker standards run alongside. However, this is possible only for DNA molecules of low complexity like plasmid DNA. For genomic DNA, especially from eukaryotes, one cannot discern a pattern of bands because the highly complex DNA gives a mixture of numerous fragments, such that only a smear is seen. To detect specific genes in the smear, the method of Southern blot analysis is applied followed by DNA hybridization with a specific probe (radioactive and non-radioactive) which can be tracked by autoradiography. Some of these methods are routinely used to screen 'ransgenic plants for the presence of the transferred genes.

Other methods include the isolation of RNA, northern blot analysis and PCR analysis. The expression of the specific gene (in a transgenic plant) can be assessed by running protein gels and carrying out Western blot analysis.

Gene cloning includes the preparation of recombinant DNA, construction of vectors and transformation of bacterial host cells with the constructs. Genetic engineering involves the transfer of these genes (cloned in this manner as constructs) to plant cells by employing a variety of methods. This chapter deals with the following:

1. Isolation of DNA
2. Isolation of RNA
3. Quantification of DNA / RNA
4. Restriction analysis of DNA
5. Southern blotting
6. DNA hybridization with radioactive probes
7. DNA hybridization with non-radioactive probes
8. Northern blotting
9. Dot and slot blotting
10. Western blotting
11. PCR amplification of DNA
12. C-DNA cloning by RT- PCR
13. RAPD analysis
14. Gene cloning and recombinant DNA techniques:
    - Isolation of the plasmid DNA (p Bluescript II KS +) from Escherichia coli strain.
    - Digestion of the plasmid DNA (vector) and the λ DNA (insert) wit the restriction endonuclease Eco RI.
    - Ligation of vector DNA and insert DNA to produce a recombinan vector.
    - Preparation of competent cells of the *E. coli* host.
    - Transformation of E. coli cells with the recombinant plasmid.
    - Selection of transformants.
    - Analysis of the recombinant plasmid by restriction digestion.

ISOLATION OF DNA

A prerequisite for molecular cloning techniques is the isolation of pure DNA free of protein and RNA contamination. There are several methods of DNA isolation and these vary according to the type of organism, the purpose for which the DNA is purified and the preferences of the individual. The CTAB method was described by Murray and Thompson (1980) and the SDS method by Dellaporta *et al.* (1983). A modified CTAB method of DNA isolation is described here. The DNA can be isolated by lysing the cells in buffers containing sodium dodecyl sulphate (SDS) or cetyl trimethyl ammonium bromide (CTAB) in the presence of EDTA, followed by further extraction with phenol and chloroform mixtures to remove proteins and other contaminants. The enzyme RNase is used to remove RNA. Depending on the extent of purity of DNA desired, proteinase K can be used to digest contaminating proteins and the RNase.

Isolation of plant genomic DNA

Plant material

Freshly harvested young leaves from healthy plants.

Reagents needed

- **DNA extraction buffer**

  2 % (w/v) CTAB

  1.5 M NaCl

  25 mM EDTA

  0.2 % (v/v) 2-mercaptoethanol

  100 mM Tris - HCl, pH 8.0
- **TE buffer**

  10 mM Tris - HCl, pH 8.0

  1 mM EDTA
- 24 : 1 chloroform / isoamyl alcohol mixture
- 3 M NaOAc
- Isopropanol
- Ethanol (100 % and 70 % v/v)

- RNase A enzyme (10 mg / ml)
- Liquid nitrogen

Protocol

1. Harvest 1 g of leaves (the leaves can be frozen in liquid nitrogen and stored at -80° C to be used whenever needed) and grind to a smooth paste with 3 ml of extraction buffer using a pre-chilled mortar and pestle.
2. Transfer the contents to Oakridge centrifuge tubes and incubate for 1 hour at 60°C.
3. Cool tube to 37° C, and add RNase to an end concentration of 20 g / ml. Incubate for 30 min at 37° C.
4. Add equal volume of the 24: 1 chloroform / isoamyl alcohol mixture. Cap the tubes and invert them twenty times.
5. Centrifuge the samples at 8,000 rpm for 5 min at room temperature.
6. Transfer the top layer into clean Oakridge tubes (do not use glass pipettes for this, as DNA sticks to glass and will be sheared). Repeat steps 4 and 5.
7. Transfer the top layer into clean tubes and add 1/10 volume 3 M NaOAc and 0.75 volume ice cold isopropanol.
8. Gently invert tubes 8 to 10 times to bring the DNA out of solution. If you see a precipitate form, you can centrifuge the DNA right away, but if none is visible, incubate at-20^0 C for 1-2 hours (can be stored for a longer time). Centrifuge at 8,500 rpm for 20 min to pellet the DNA.
9. Discard the supernatant. Wash pellet with 70% ethanol and airdry the pellet. Suspend the DNA pellet in 300 µl of TE buffer. After the pellet dissolves, transfer to 1.5 ml eppendorf tubes. Wash down the empty tubes with 100 µl of TE buffer and add to the samples.
10. Quantify (explained in later pages) the amount of DNA and store it at -20^0 C.

Isolation and purification of plasmid DNA

Plasmids are essential vectors used in gene subcloning, sequencing and expression. The extraction and purification of plasmid DNA is a very important technique in molecular biology. Plasmids are purified from liquid bacterial cultures grown by inoculating a single bacterial colony in appropriate medium. Most of the plasmids are high copy number vectors and replicate up to 100 to 300 copies per cell so that they result in a high yield.

The protocol given below is modified from the protocols published by Birnboim and Doly (1979). It allows for the purification of small to large amounts of plasmid DNA without resorting to the banding in CsCl gradients. The mini-preparation (M) and the large scale preparation (L) of the plasmids is explained together.

Reagents

- **LB (Luria-Bertaini) broth**

  10 g Bacto-tryptone

  5 g Bacto-yeast extract

  5 g NaCl

  Bring up to 1 liter with distilled water and adjust the pH to 7.5. Autoclave the medium, and add the required antibiotics after it has cooled below 50^0 C.The same medium can be made for petriplates by the addition of 15 g agar / l prior to autoclaving.

- **E buffer (pH 8.0)**

  10mM Tris - HCl

  1 mM EDTA

- **STE buffer (pH 8.0)**

  100 mM NaCl

  10 mM Tris - HCl

  1 mM EDTA

- **TGE buffer (lysis buffer, pH 8.0)**

  50 mM Glucose

  25 mM Tris - HCl

  10 mM EDTA

  Add 5 mg / ml Lysozyme before use.

- **Alkaline solution**

  0.2 N NaOH

  1 % SDS

- 7.5 M ammonium acetate, pH 7.6
- 2.0 M ammonium acetate, pH 7.4
- PCIA (25:24:1 of phenol / chloroform / isoamyl alcohol)
- RNase - A (DNase - free) enzyme
- Isopropanol
- Ethanol (100 % and 70 %)

Growing bacterial cultures

For the large scale method, inoculate a single colony or 50 μl of previously frozen cells containing the plasmid of interest in 1000 ml of LB medium containing the appropriate antibiotics (e.g. 50μg / ml ampicillin.) depending on the specific anti biotic - resistant gene carried by the specific plasmid.For the mini - preparation, inoculate a single colony or 10 μl of previously frozen cells containing the plasmid of interest in 2 ml of LB medium and the appropriate antibiotics . Culture the bacteria at 37^0 C overnight (22-24 hours) with vigorous shaking (250 rpm).

Protocol

The methods used for both the large scale and mini- preparation are the same and are designated as (L) and (M) wherever neccessary. The mini-preps are sized for working with microfuge tubes and microfuges, which usually reach a maximum speed of 12,500 rpm.

1. Collect the bacterial culture in tubes and centrifuge (L: 5000 rpm / 5 min ; M: 12,500 rpm / 2 min). Discard supernatant.
2. Wash the pellet with STE buffer (L : 50 ml : M : 200 μl).
3. Resuspend pellet in TGE buffer (L : 50 ml : M : 200 μl). Vortex and incubate at room temperature for 5 min.
4. Add a freshly prepared alkaline solution (L:100 ml ; M: 400 μl). Cap the tubes and mix well by inverting five times. Incubate for 5 min on an ice-water bath.
5. Add ice cold 7.5 M ammonium acetate (L: 75 ml ; M: 300 μl), mix by gently inverting the tubes five times and incubate the tubes on an ice-water bath for five min.
6. Centrifuge the contents at 12,500 rpm for 10 min at 40° C and transfer the clear supernatant to clean tubes. Avoid the particulate material on

the top and the loose pellet on the bottom. Add 0.6 volumes of isopropanol and incubate at room temperature for10 min.

7. Centrifuge mixture at 12,500 rpm for 10 min at 40° C and discard the supernatant. Add 2 M ammonium acetate (L: 25 ml ; M: 100 μl) to the pellet, vortex and incubate in an ice bath for 5 min.
8. Centrifuge mixture at 12,500 rpm for 10 min at 40° C and save the supernatant. Add isopropanol (L: 25 ml ; M: 100 μl)and incubate at room temperature for 10 min.
9. Centrifuge mixture at 12,500 rpm for 10 min at 40° C and discard the supernatant. Wash the pellet with 70 % ethanol and discard the ethanol after centrifuging the contents. Air dry the pellet and dissolve it in TE buffer (L: 5 ml; M: 50 μl). Optional : Add RNase at a concentration of 20 μg /ml and incubate at 37° C for 30 min.
10. Add PCIA (L: 5 ml; M:50 μl) to the contents and mix by gently inverting the tube 5 times. Centrifuge mixture at 12,500 rpm for 10 min at 40° C. Collect the top layer carefully with a pipette. Repeat the PCIA treatment again and collect the clear upper layer.
11. Add 7.5 M ammonium acetate (L : 2.5 ml; M: 25 μl) and isopropanol (L : 7.5 ml ; M: 75 μl). Mix and incubate at room temperature for 10 min. Centrifuge (L : 8000 rpm, 20 min ; M : 12,500 rpm, 10 min) at 4° C and save the DNA pellet.Wash the pellet twice with 70 % ethanol and air dry it.
12. Resuspend the pellet in TE buffer (L: 2 ml; M: 25 μl). Quantify the amount of DNA and store at – 20° C.

ISOLATION OF RNA FROM PLANT TISSUE

RNA molecules (compared with DNA molecules) are very unstable due to the abundance of RNases in biological materials. This makes the procedure of RNA isolation more complicated. RNases are present everywhere including our hands and bacteria and fungi on airborne dust particles. To prevent RNA contamination, gloves should be worn at all times and changed frequently. All glassware should be treated with 0.1 % diethyl pyrocarbonate (DEPC) and autoclaved before use and whenever possible, disposable plasticware should also be autoclaved before use. Glassware should be should be baked at 250° C overnight. Disposable and sterile polypropylene centrifuge tubes are recommended for the RNA isolation. The procedure involves the use of guanidine thiocyanate (Gilsin *et al.* 1974; Ullrich *et al.* 1977). Guanidine thiocyanate and β - mercapto ethanol are specifically included in the extraction buffer to prevent the endogenous RNase activity. Whenever possible the solutions should be treated with 0.05 % DEPC at 37^0 C for 2 hrs or overnight at room temperature and then autoclaved for 30 min to remove the DEPC. The solutions like tris buffers which cannot be treated with DEPC should be prepared with DEPC-treated distilled water.

Caution !: Guanidium thiocyanate is a toxic substance.

Plant material

Freshly harvested leaves ground to a fine powder with liquid nitrogen.

Reagents needed

- **GTC buffer**

  4 M guanidium thiocyanate

  25 mM sodium citrate (pH 7.0)

  0.5 % sarkosyl (w/v)

  0.1 M β - mercaptoethanol

- **STE buffer**

  0.5 % SDS (w/v)

  10mM Tris - HCl (pH 7.5)

  1mM EDTA

2 M sodium acetate

Water saturated phenol

Thaw crystals of phenol at 65° C in a water bath and mix one part of phenol and one part of sterile distilled water. Mix well and allow two phases to separate. Store at 4° C.

4 M lithium chloride

Chloroform / isoamyl alcohol mixture (24 : 1)

Liquid nitrogen

Protocol

1. Take 2 g of the finely ground tissue in a sterile polypropylene tube and add 5 ml of GTC buffer. Vortex well.
2. Add sequentially: 0.5 ml of 2 M sodium acetate, 5 ml of phenol and 1 ml of chloroform. Vortex well.
3. Centrifuge at 10,000 rpm for 10 min at 4° C and transfer the upper phase to another tube.
4. Add 5 ml isopropanol and allow to precipitate at -20^0 C for 1.5 to 2 hrs and centrifuge at 10,000 rpm for 20 min at 4^0 C.
5. Resuspend the pellet in 1.5 ml of 4 M lithium chloride and transfer to microfuge tubes. Spin for 10 min in a microfuge to sediment the RNA.
6. Repeat step 5. Resuspend the pellet in 750 µl of STE buffer. Add 750 µl of the chloroform / isoamyl alcohol mixture and vortex. Spin for 10 min and collect only the upper phase. Repeat the extraction with the chloroform / isoamyl alcohol mixture.
7. To the clear upper phase, add 2 M sodium acetate (1: 10 v/v) and 600 µl isopropanol. Incubate at -20^0 C for 2 hrs or overnight.
8. Centrifuge at 10,000 rpm for 20 min at 4^0 C and retain the RNA pellet. Rinse the RNA pellet with 75 % ethanol and dry the pellet under vacuum for about 15 min.
9. Dissolve the total RNA in 200 µl of sterile - DEPC treated water. Quantitate the RNA and check the quality by 1 % agarose, formaldehyde denaturing gel electrophoresis. Store the RNA at -20^0C for future use. This RNA can be used for Northern blot analysis, or dot blot hybridization.

QUANTIFICATION OF DNA AND RNA

The most popular method of DNA / RNA quantification involves the use of a UV spectrophotometer. Spectrophotometric measurement indicates the amount of ultraviolet irradiation absorbed by the bases of the nucleic acid (Sambrook *et al.* 1989).

Procedure

1. Measure 10 µl of the DNA sample into an microcentrifuge tube and dilute it with 990 µl of T E buffer (this amounts to a 100 fold dilution)

In the case of RNA, dilute 10 μl of each sample to 990 μl of sterile - DEPC treated water.

2. Record the absorbance of the sample at 260 and 280 nm after checking with appropriate blanks comprising the TE buffer.

3. Calculate the ratio between the readings at 260 and 280 nm to provide an estimate of the purity of the nucleic acid. Pure samples of DNA have a value of 1.8 and those of RNA have a value of 2.0.

4. Calculate the DNA or RNA concentration as follows: DNA/RNA concentration) μg/ml = (dilution factor) × (the average of duplicate absorbance readings at 260 nm) × (50 μg DNA or 40 μg RNA per absorbance unit at 260 nm). An OD of 1 corresponds with approximately 50 μg/ml of double - stranded DNA, 40 μg/ml of single stranded DNA and RNA and ~20 μg/ml for single stranded oligonucleotides.

5. After the measurement of the concentration of the DNA/RNA, appropriate dilutions can be made to represent μg / μl.

RESTRICTION ANALYSIS OF DNA

To get an idea of the general map of the isolated DNA and to characterize it, a restriction analysis could be performed by digesting it with restriction endonucleases. However, the restriction analysis of the high molecular weight plant genomic DNA will only yield a smear composed of numerous fragments. To detect a particular fragment from this complex pattern,it is essential to carry out Southern blot analysis (which will be dealt with later). In case of small, homogeneous DNA like the plasmid DNA, it is possible to measure the fragments after the restriction analysis and to construct a map.

Restriction endonucleases are prepared from bacteria. They recognize and cleave specific sequences of DNA. Each enzyme has a particular recognition sequence composed of 4 to 12 or more bases. Most of the recognition sequences are palindromic (the sequence reads the same going forward and backward on the top and bottom strands in the 5 to 3 directions. A restriction enzyme cleaves the phosphate - sugar backbone of the DNA at the recognition sequence or near it. Blunt or sticky ends are produced as a result and the sticky ends make it possible for the DNA molecules to be joined to one another, to create recombinant molecules. There are atleast 200 enzymes available commercially and each enzyme needs specific conditions of temperature, salt concentration and pH which are taken care of by the buffers supplied by the manufacturers. If the DNA molecule has one or more sites for a restriction enzyme, the DNA molecule will be cut into fragments which can be separated according to their size by gel electrophoresis. There are many types of apparatus used for electrophoresis including the vertical and horizontal types.

Gel electrophoresis involves the use of a gel matrix composed of agarose or polyacrylamide. A voltage potential is applied by means of a power supply. The electrode at which the electrons enter the gel apparatus from the power supply (black or blue) is called the cathode and is negatively charged, whereas the electrode at which electrons leave the box and re-enter the power supply (red) is called the anode and is poitively charged. The flow of electrons set up a potential energy difference between the electrodes. The potential is measured in volts and the current generated is measured in milliamperes. The voltage potential induces the negatively charged molecules (DNA or RNA loaded in wells) to move through the gel to the positive electrode.

Several factors affect the migration of nucleic acids in a gel. They are:

1. The size of the molecule (represented as base pairs in case of DNA or nucleotides in case of RNA): smaller molecules migrate faster.
2. The conformation of the molecule: DNA is supercoiled, circular or linear and DNA has varying degrees of secondary nature (such as tRNA). The supercoiled molecules and those with secondary structure migrate faster.
3. The density of the gel matrix: Molecules migrate very slowly in a high density gel which has smaller pores.
4. The voltage applied: Molecules move faster at high voltage.
5. The type of buffer used and its ionic strength.

Protocol

Overview of restriction analysis:

- Digestion of DNA.
- Preparation of agarose gel for electrophoresis.

- Electrophoresis of the digested DNA.
- Viewing the gel and photographing it.
- Analysis of the gel.
- Purification of DNA fragments from agarose gels.

Digestion of DNA

DNA to be digested

Plant genomic DNA (extracted from any plant).

Plasmid DNA (any known recombinant plasmid construct or vector).

Use 1-3 μg of DNA per digestion.

Restriction enzymes

For the plant DNA, use either Eco R1 or Hind III.

For the plasmid, work with two digests :

- the specific enzyme that cuts at the unique site to linearize it.
- the specific enzyme that isolates a particular gene in the construct.

Restriction enzymes are supplied along with their buffers. Store the enzymes at -20° C.

1 μl of enzyme is enough to digest 10 μg of DNA in an hour.

1 unit of enzyme is defined as the amount required to completely digest 1 μg of DNA in 1 hr in the recommended buffer in a 20 μl reaction.

Protocol

1. Set up the restriction digestion reactions in microfuge tubes as follows:

X μl	DNA solution (1-3 μg DNA)
2 μl	buffer (specific to the restriction enzyme used)
0.5	μl restriction enzyme
y μl	sterile distilled water
20 μl	**total volume**

2. Include undigested DNA controls with x μl DNA and y ul of sterile water and buffer to make up a total volume of 20 μl.

3. Use a new pipette tip for each reagent. Make sure the pipette tip is touching the side of the tube when the reagents are added.

4. For setting up the digestion reactions, mix the water, buffer and DNA before addding the enzymes. Tap gently for the contents to mix well.

5. Label each tube accurately.

6. Incubate the reactions at 37° C for 1 hour (or a different temperature if specified for the restriction enzyme). Proceed with the electrophoresis of the samples on an agarose gel.

Preparation of agarose gel for electrophoresis

The DNA fragments created as a result of the restriction digestion can be separated according to their size by electrophoresis. Polyacrylamide gel electrophoresis or agarose gel electrophoresis can be used for the restriction analysis but the latter is generally preferred. As a starting point, 0.8 % agarose is generally suitable. After electrophoresis, the nucleic acids in the gel are stained with a fluorescent dye, ethidium bromide (EtBr), which permits the visualization of the DNA or RNA in the samples under UV illumination . DNA binds the EtBr well but RNA binds it only in regions that form double stranded structures. EtBr can be added to the gel solution so there is no need to stain it later.

Caution ! Et Br is a mutagen. Use gloves !!

Materials needed for the preparation of 0.8 % agarose gel:

- Agarose (molecular biology grade)
- **Tris - borate electrophoresis buffer (TBE) 5 x**

  (Concentrated stock solution per litre)

  54 g Tris base

  27.5 g boric acid

  20 ml 0.5 M EDTA (pH 8.0)

  Prepare a 1:10 dilution of the stock to obtain 0.5 x TBE.

- **Ethidium bromide (EtBr)**

  Prepare a stock solution of 10 mg / ml and use 2 µl for 100 ml of buffer .

Procedure

Prepare a clean electrophoresis gel mold by sealing both the open edges with tape. Place a gel comb in the slot. Prepare the 0.8 % agarose gel solution by melting agarose in the 0.5 X TBE and heating it until it boils. Swirl to mix and let it cool to 45° C. Add Et Br, swirl to mix and pour in the prepared mold with the comb intact. Wait about an hour for the agarose to solidify (it produces a uniformly thin layer). Remove the comb carefully without distorting the wells before proceeding with the loading of samples and electrophoresis.

Electrophoresis of the digested DNA

The samples of nucleic acids that need to be loaded on the gel for electrophoresis are mixed with a tracking dye mixture (loading dye) and loaded into the wells. These buffers increase the density of the sample ensuring that it drops evenly into the well, add color to the sample,and contain dyes that move toward the anode (in an electric field) at predictable rates. They also provide a visible moving front of the electrophoresis. The following loading buffer is 5 X.

50 % glycerol

0.1 % bromophenol blue

0.1 % xylene cyanol

(Use enough loading dye per sample to make it 1X.)

Procedure

Keep all the samples ready. Load a DNA size marker standard composed of a commercially available lambda digest to measure the size of the restriction fragments that separate out. Add 2 µl of the loading dye to each sample (so that the loading dye is at least 1X concentration). Tap well to mix. Remove the tape from the ends of the gel plate and set it in the electrophoresis unit (remove the comb carefully without distorting the wells). Pour the TBE buffer to fill the tank and cover the gel completely. Load 20 µl of sample into each well. Plan the lanes as follows:

DNA size marker

- uncut plant genomic DNA
- digested plant genomic DNA uncut plasmid DNA
- plasmid DNA digested with enzyme I
- plasmid DNA digested with enzyme II
- DNA size marker

Cover the unit and connect it to the power supply and run the gel at approximately 100 v until the lower dye has run about 3/4 length of the gel. Small gel box units need less voltage to achieve the same effective current. Disconnect the unit and gently lift (wear gloves) the gel from the plate and place it on the transilluminator.

Viewing the gel and photographing it

View the gel under UV light (**Caution! Use UV protective goggles or face mask**). Photograph the gel with a polaroid film under UV light by placing a transparent ruler so that the zero aligns with the wells. The characteristics of the migrated samples can then be analysed and the size of the restriction fragments (visualized as bands) confirmed in relation to the DNA size marker standard bands (Fig - 26).

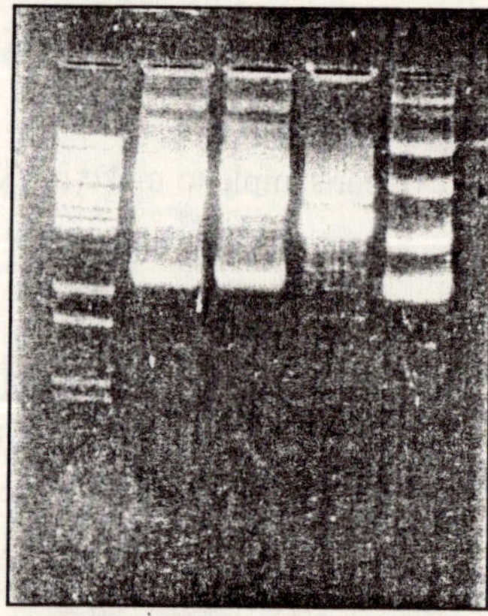

Fig. 25. Restriction analysis of DNA (gel with bands).

Analysis of the gel

The uncut plant genomic DNA which is of high molecular weight, collects as a thick band near the well and does not migrate very far. The digested plant genomic DNA however forms a smear composed of innumerable fragments. The uncut plasmid DNA migrates in the gel as three bands, composed of open/circular (top band), linear (middle) and closed circular (lowest band). The plasmid DNA digested with the enzyme that cuts at the unique site will linearize the plasmid andit migrates downwards according to its size. The plasmid DNA digested with another enzyme results in fragments that migrate according to their sizes. The sizes of all these fragments can be confirmed in relation to the DNA size markers. With the help of the data from restriction analysis, a restriction map of the plasmid can be prepared. If the plasmid has a unique site for enzyme A, the total size of the plasmid can be determined easily. If the enzyme B cuts the plasmid to create two fragments (with one comprising the recombinant part and the other representing the rest of the plasmid), then a restriction map showing a circular DNA with the restriction sites and sizes of the fragments can be prepared. If more enzymes are used, a more accurate map can be constructed. Under some circumstances, double digestions may be desirable. This method is also used for the confirmation of the presence of a specific gene in the plasmid. After the confirmation of the specific fragment, it is possible to elute and purify it from the gel.

Purification of DNA fragments from agarose gels

Specific DNA fragments can be isolated from the gel after confirmation (comparing the sizes of the bands with the DNA size markers). Isolation of specific fragments may be necessary for the preparation of DNA probes to be used in the DNA or RNA hybridization. Specific DNA inserts can also be separated by restriction digestion and electrophoresis and after elution, ligated to new vectors for subcloning. A simple method for the recovery of DNA from after electrophoretic separation in low gelling temperature agarose gels was reported by Wieslander (1979) and Parker and Seed (1980). Two methods of elution of DNA modified from the above are described here.

Procedure 1: Elution of DNA fragments by gel slicing and melting

Requirements

- T E buffer
- 3 M sodium acetate buffer (pH 5.2)
- 1 % agarose gel prepared with low-melting agarose (add Et Br)(explained earlier)
- Water saturated n-butanol
- 70 % and 100 % ethanol

Procedure

1. Electrophorese the digested DNA on the agarose gel. Locate the specific band by screening the gel placed on the transilluminator and excise the band with a sterile razor blade as neatly as possible (use a plastic wrap below the gel to protect the surface of the transilluminator from the razor blade). Cut the gel strip into tiny pieces.

2. Place the gel slices in microcentrifuge tubes and add two volumes of TE buffer.

3. Melt the gel slices by incubating the tubes in a 70° C water bath. Chill the melted gel solution immediately on ice. Incubate the microcentrifuge tubes at -70° C for 20 min.

4. Leave the microcentrifuge tubes at room temperature to thaw. Centrifuge at 10,000 rpm for 5 min at room temperature. Remove the liquid phase containing the DNA into clean tubes. This can be used directly for labeling or the DNA can be purified to remove the Et Br by extraction with three volumes of water-saturated butanol and precipitated.

5. Add 0.1 volume of 3 M sodium acetate buffer (pH 5.2) and 2.5 volumes of chilled 100 % ethanol to the DNA solution and incubate at - 70^0 C for 1 hr to precipitate the DNA.

6. Centrifuge the contents at 12,000 rpm for 5 min at room temperature. Discard the supernatant and wash the DNA pellet with 1 ml of 70 % ethanol. Air dry the pellet and dissolve in an appropriate volume of TE

buffer or distilled water. Measure the concentration of the DNA and store it at -20^0 C.

Procedure 2: Elution of DNA fragment from agarose gels (crush method)

Requirements

- **Tris- saturated phenol**

  Prepare water saturated phenol (thaw phenols crystals at 65° C in a water bath and mix one part of phenol and one part of sterile distilled water. Mix well and allow two phases to separate.). Take an aliquot of the water - saturated phenol and add an equal volume of 0.5 M Tris-HCl (pH 8.0). Mix well and let the phases separate at 15 to 30° C. Transfer the upper aqueous phase to another tube. Add equal volumes of 0.1 M Tris-HCl (pH 8.0) and repeat the above step. Check the pH of the aqueous phase and repeat until the pH is 8.0.
- Phenol / phenol-chloroform:
- Glycogen
- 5M ammonium acetate
- Isopropanol / ethanol
- 70 % ethanol
- TE buffer

Procedure

1. Locate the band in the agarose gel and cut it out with a sterile razor blade. Transfer the gel to a microfuge.

2. Photograph the gel for reference.

3. Add 1 μl glycogen, 100 μ l5M ammonium acetate and 1 volume of isopropanol or 2 volumes of ethanol.

4. Incubate at -20° C, overnight or at -70° C, for 20 min. Spin at 40° C for 30 min at 12,000 rpm.

5. Wash the pellet with 70 % ethanol and dissolve it in TE buffer.

Procedure 3: Recovery of DNA from agarose gel by electroelution

Requirements

- Electroelution apparatus
- Syringe
- 5 M potassium acetate
- 0.5 X TBE buffer (composition given earlier)

Procedure

1. Locate the specific band by observing its position in the gel placed on the transilluminator and excise the band with a sterile razor blade as neatly as possible. Cut the gel strip into tiny pieces.

2. Fill each of the tubes in the elutor kit with the 5 M potassium acetate by using a syringe (take care to avoid air bubbles). Place the gel slices in the cavities. Fill the tank with the 0.5 X TBE buffer and let it cover the gel slices.

3. Plug in and run the elutor for 15 min or more at 50 volts. Check a sample gel slice for total elution of the DNA (on the transilluminator) and stop the run.

4. Gently remove the potassium acetate (with the DNA dissolved in it) from the tubes using a syringe and precipitate the DNA (explained in earlier procedure).

5. Purify the DNA by washing in 70% ethanol. Dissolve the DNA in TE and check the concentration. Store the eluted DNA sample at - 20° C.

SOUTHERN BLOTTING

Southern blotting is one of the most important techniques of molecular biology. Developed by Southern (1975), this procedure allows the detection of a particular sequence of DNA that is only a small percentage of a complex mixture of DNA fragments (e.g., the smear observed after restriction analysis of the plant genomic DNA). The detection of DNA by hybridization involves transfer of the DNA from the gel onto nitrocellulose or nylon membrane. The transfer or blot occurs by capillary transfer. Quicker and efficient transfer of the DNA is possible by further breaking up the DNA fragments by the process of depurination. Alkaline salt denaturation of the double stranded DNA, followed by neutralization is carried out before the blot to facilitate efficient transfer of the DNA and also binding to the membrane.

The whole process of Southern transfer and hybridization involves four steps: Digestion of the DNA, gel electrophoresis of the fragments, transfer of the DNA to a membrane, hybridization of the DNA with the probe and autoradiography to detect the signal. Though the general principles remain the same, certain variations may exist in the protocols used by different laboratories. The buffer is drawn through the gel and through the membrane before being absorbed by stacks of blotting papers or tissues placed above by capillary action. This causes the DNA fragments to be transferred onto the nitrocellulose membrane. Several vacublot units have been created to speed up the process of Southern blotting.

Materials

- Nylon or nitrocellulose membranes.
- Baking oven.
- 3 M Whatman paper
- Blotting tray
- Plate and 4 test tube caps to serve as columns or a plexiglass blotting plate with legs
- Saran wrap
- Paper towels
- Any object like a bottle with 500 ml of water to serve as a weight

Solutions

- **Depurination solution**

  0.25 N HCl

  Make up 10 ml of concentrated HCl to 500 ml with dH_2O to prepare 0.25 N HCl

- **Denaturation solution:**

  1.5 M NaCl + 0.5M NaOH

  Dissolve 43.8 g NaCl and 10g NaOH in dH_2O. Make up to 500 ml and autoclave

- **Neutralizing solution:**

  1M Tris pH 7.4 and 2M Nacl

  Dissolve 60.55 g Tris in dH_2O and adjust pH to 7.4. Dissolve 58 g NaCl and make upto 500 ml with dH_2O and autoclave.

  Transfer (blotting) solution

  10x SSPE buffer

  1.8 M sodium chloride

  0.1 M sodium phosphate (dibasic)

  0.01 M EDTA

  Bring upto 1 liter with distilled water and adjust pH to 7.7

Procedure

1. Digest plant genomic DNA with one or two (double digest) restriction enzymes as instructed in the earlier experiment and run the run the gel to separate the fragments (this usually appears as a smear). Run appropriate controls which can be the DNA from some other source which does not contain the gene of interest and a 1 : 1000 dilution of digested plasmid DNA carrying the DNA that will be used as the probe (bands should not be visible.on the EtBr stained gel, or it will act as a sink in binding the probe.

2. Photograph the gel for record's sake.

3. Place the gel in a plastic tray and soak it in 0.25 N HCl with slow, gentle shaking at 10-20 rpm for 10 min.

4. Drain off the HCl, rinse with distilled water and add the denaturation solution and continue treatment for 15 min at room teperature. Repeat denaturation with fresh solution.

5. Drain off the denaturation solution, gently rinse the gel with distilled water and add the neutralizing solution. Continue the treatment for 15 min. Repeat the neutralization. Cut the nitrocellulose membrane/nylon membrane to be the exact size of the gel to be transfered, and one sheet of 3 M Whatman filter paper. Cut out another larger piece of 3 M Whatman filter paper. Handle the nitrocellulose paper with gloves and mark it on the upper left or right corner with a pencil.

6. Rinse the membrane and moisten the filter papers for 5 minutes in distilled water and later soak them in the transfer solution for 20 min.8. Take the blotting tray and place a large sponge in it. Pour the transfer solution and saturate the sponge with it. Lay the bigger Whatman (pre wet) paper over the sponge. Alternatively, place the large Whatman filter paper on a flat elevated plate in such a way that the two ends hang into the transfer buffer and serve as wicks. Fill the blotting tray with the transfer buffer to the edge of the flat plate but do not allow the solution to come up over the plate ! (refer Fig - 11)
7. Place the gel carefully on top of the filter paper starting from one side and with the well-side facing down. Run a pipet or glass rod gently over the gel to remove any air bubbles. Cover the rest of the area all around the gel with parafilm to prevent contact of the top paper towels with the wick.
8. Place the nitrocellulose membrane over the gel. The marked side should face the gel. Take care to prevent any air bubbles in between the gel and the membrane as they will interfere with the transfer.
9. Gently overlay the membrane with an exactly sized Whatman filter paper. Place a stack of paper towels on top of the gel assembly and put a light weight on top of the papers.
10. Leave over night for the transfer to be completed by capillary action. Remove the paper towels and the membrane filter and use a ball point pen to trace the position of the wells on the membrane before removing it from the gel.
11. Air-dry the membrane and bake it at 80° C for about two hours in a vacuum oven, or overnight in a regular oven.
12. Store the membrane in a sealed polythene bag at 4° C and prepare for the hybridization procedure.

DNA HYBRIDIZATION WITH RADIOACTIVE PROBES

Hybridization of the DNA on the membrane with specific probes is essential to study RFLPS. DNA hybridization with radioactive probes involves four steps:

1. Isolation of the DNA sequence to be used as the probe

2. DNA labeling
3. DNA hybridization
4. Autoradiography.

Isolation of the DNA sequence to be used as the probe: (from the plasmid construct)

This procedure is carried out in a situation where a foreign gene construct (such as a recombinant plasmid vector) is transferred to plant tissue (genetic engineering) which is then cultured *in vitro* to produce a transgenic plant. The Southern blotting of the genomic DNA is followed by hybridization with a probe specific to the foreign gene (in the construct) that was transferred into the plant tissue. This probe can be isolated from the recombinant plasmid constructs. Recombinant plasmids (DNA) should be isolated from the bacteria harbouring them. For the random primer labeling described below, the DNA fragment (insert) in the plasmid should be freed by digestion with specific restriction enzymes and isolated by gel purification as previously described.

Procedure

- Grow bacteria in LB medium as explained earlier and isolate the plasmid DNA.
- Digest the plasmid DNA for 1 hour with suitable restriction enzymes.
- Check for the presence of the fragment (insert) by agarose gel electrophoreσis and elute the insert DNA (refer to the protocol explained earlier).
- Label the DNA and use it as a probe

DNA labeling: (Preparation of radioactive probes)

There are several methods for labelling DNA fragments. The choice of the method depends on the sensitivity needed. Radioactive labeling (especially $_{32}$P labeled compounds) remains the method of choice for the generation of highly sensitive probes but efforts have been made in recent years to develop methods of non-radioactive labeling.

The method for labelling DNA with random oligonucleotides (called oligolabelling) was developed by Feinberg and Vogelstein (1983 and 1984).

A DNA template is taken and a commercially available set of 6 base random oligomers are added as primers to initiate the synthesis of the complimentary DNA strand by the DNA polymerase. The T7 DNA polymerase has an advantage over other DNA polymerases in labelling very small quantities of DNA (Arrand, 1985 and Prober *et al.* 1987). This is supplied with the labelling kit of Pharmacia Biotech and generates a highly labelled DNA.

Materials

- Probe DNA (25 to 50 ng)
- **The labeling kit from Pharmacia Biotech** [T7 Quick prime kit]

  It contains The T 7 DNA polymerase enzyme, the primer buffer, the dNTPs (mixture of dGTP, dTTP and dATP)], random oligodeoxyribonucleotides and the labeled nucleotide,[alpha ^{32}P] - dCTP (3000 C i / mmol) 50 μCi is actually needed for 25 ng of DNA.
- Distilled water

Procedure

1. Bring the probe DNA (isolated insert) to 34 μl with TE in a microfuge tube. Heat (boil) in a 100° C heat block or water bath for 10 min to denature the DNA.
2. Add 10 μl reagent mix, 5 μl of [^{32}P]- dCTP and 49 μl of distilled water. Add 1 μl T7 DNA polymerase. Tap gently and centrifuge (pulse) to collect the contents together.
3. Keep this reaction mixture for 20 min at 37° C.
4. Use this probe solution for hybridizing the denatured DNA on the membrane filters after denaturing it by heating at 95-100° C for 2 min. Cool immediately on ice and proceed.
5. Plug the bottom of a 1 ml disposable syringe with a little bit of glass wool and put the syringe in a 15 ml disposable plastic centrifuge tube. Make a slurry with Sephadex G-50 in STE buffer (TE + 100 mM NaCl + 0.02 % NaN_3, to prevent bacterial contamination), pack it into the syringe and spin at top speed in a clinical centrifuge for 2 min. Pack more Sephadex till it is almost 0.9 ml. Place the syringe assembly with the end extending into an eppendorf tube in another 15 ml tube.

6. Bring up the probe mix to 100 µl with STE and layer it gently onto the top of the column. Cover with a piece of parafilm and spin again for 2 min.

7. Wash out the probe tube with STE and layer it onto the column and spin again. Transfer the collected purified probe to a new, sterile microfuge tube.

8. Check the specific activity of the mixture by measuring 2 µl of it with a scintillation counter (optimum counts should be I - 10×10^8 CM/µg template). Dispose of the waste appropriately.

DNA hybridization

After baking, the membrane filter can be subjected to prehybridization followed by hybridization and autoradiography.

Reagents

- **20 X SSC solution**

  3 M Sodium chloride

  0.3 M sodium citrate (trisodium citrate)

  Bring it upto 1 L with distilled water, adjust the pH to 7.0 and autoclave.

  5 X SSC solution

  Dilute 20 X SSC solution four times with sterile distilled water.

  2 X SSC solution

  Dilute the 20 X SSC solution ten times.

- **50 X Denhardt's solution**

  1 % (w/v) bovine serum albumin

  1 % (w/v) Ficoll (type 400, Pharmacia.)

  1 % (w/v) poly vinyl pyrrolidone (PVP)

  Dissolve one after the other and adjust the final volume to 500 ml with sterile distilled water. Divide into 50 ml aliquots and store at -20^{0}C. Dilute 10 fold into prehybridization and hybridization buffers.

- **Prehybridization buffer**

  5 X SSC

  0.5 % SDS

  5 X Denhardt's solution

  0.2 % denatured and sheared salmon sperm DNA (explained below)

- **Hybridization buffer**
  5 X SSC
  0.5 % SDS
  5 X Denhardt's solution
  0.2 % denatured salmon sperm DNA
  ^{32}P - labeled DNA probe* (to be added at the appropriate time).
- **Preparation of denatured salmon sperm DNA**
  Dissolve the DNA in sterile distilled water and shear it by sonication or alternatively, stir for 2 hrs or overnight to break up the DNA into small fragments.

Procedure

(Carry out all the procedures in the radioactive hood behind the shield and when using the shaker outside, cover the tray and bring it out. Wear a lab coat and gloves at all times.)

1. Take out the stored membrane filter and place it in a plastic tray or a heat sealable bag. Immerse it in 5 X SSC buffer for 5 min at room temperature.
2. Pour off the 5 X SSC and add enough prehybridization solution to cover the membrane completely. Carry out prehybridization at 42° C for 2 - 4 hrs with mild shaking at 60 rpm.
3. Denature the labeled probe by heating it in a water bath at 90^0C for 10 min (be sure to puncture the lid of the microfuge tube with a needle so that it will not pop open during the heating). Immediately chill on ice for 5 min. Briefly spin down in a microcentrifuge prior to use.
4. Pour off the prehybridization buffer and pipette the hybridization solution into a disposable tube or tray (use about 10 ml for a medium size blot). Add the probe to the hybridization buffer. Allow hybridization to proceed overnight at 42° C.

---

*The ^{32}P - labeled DNA is dangerous. A lab coat and gloves should be worn at all times and the waste substances and liquid contaminated with the isotope should be collected in labeled containers for proper disposal. The area should be surveyed for any spills and cleaned thoroughly.

5. Discard the used probe solution and the wash waste according to radioactive safety containment procedures. Wash the filter in 2 X SSC with 0.2 % SDS at room temperature with shaking at 60 rpm for 15 - 20 min. The first wash should also be collected as radioactive waste. Repeat once.
6. Wash with the same solution at 50°- 55° C with shaking for 20 min. Repeat twice. **Note:** *Varying the amount of SDS and SSC concentration, as well as the temperature will affect the stringency of the hybridization. More salt and more SDS wil denature less stringency hybridizations.*
7. Airdry the filter at room temperature for 40 min and proceed with autoradiography.

Autoradiography and X - ray film developing

Place the filter in plastic wrap and place in an exposure film cassette. In a dark room, with the safe light on, cover the filter with a piece of X- ray film and place the cassette in the dark at -80^{0}C for 1-3 days.. If an intensifying screen is available, it will reduce the exposure time four fold, but to function correctly, the cassette must be stored for 1-3 days prior to their being developed.

X ray Film development

Requirements

- Developer (e.g Kodak GBX)
- Fixer (e,g Kodak GBX)
- Developing trays

Procedure

- Remove the film cassette from the freezer and take it to the dark room.
- Remove the X - ray film from the cassette and place it in a tray containing the X - ray developer for 2-3 minutes.
- Transfer the film to another tray with water. Rinse for 30 seconds.
- Transfer the film to another tray with the fixer. Incubate it for 3 minutes.
- Wash the film in water for 2 minutes.

- Air-dry the film and mark the locations of the lanes by looking at the blot.

Automatic film development: There are several kinds of automatic film development machines. The machine is generally located in a dark room provided with a safe light (red light). The film is removed from the cassette in darkness and fed into the machine by following the instructions of the manufacturer. After the film is developed, it emerges automatically from the machine. The film can then be viewed under a regular light or on a white light box. The place where the DNA has hybridized will be darkened as a band indicating the sequences that are homologous to the probe.

DNA HYBRIDIZATION WITH NON – RADIOACTIVE PROBES

This procedure (modified from Kricka, 1992, Langer *et al.* 1992 and Leary *et al.* 1983) involves the following steps:

- Isolation of the DNA sequence to be used as the probe (follow the procedure described under the DNA hybridization with the radioactive probe).
- Preparation of non-radioactive probes (labeling).
- DNA hybridization and detection of hybridized probe.

The alternative to radioactive labeling is the non-radioactive labeling method. There are several such methods available, of which the biotin labelling of DNA by nick translation is described here. The labeling kit can be purchased from GIBCO/BRL (Life technologies Inc). After hybridization of the filters with the biotinylated probe, the DNA - DNA hybridization can be detected by using a detection system consisting of streptavidin and biotin-conjugated alkaline phosphatase. Addition of a colorless substrate for alkaline phosphatase will result in the formation of the colored product, which appears as bands.

Requirements (for all the steps)

- **BioNick labeling system (GIBCO / BRL)**

  Control DNA (5 µg of pBR 322 plasmid DNA in 0.1 mM EDTA, 10mM Tris -HCl (pH 8.0).

dNTPs 10 X

- **0.2 mM concentration of dTTP, DGTP, dCTP**

  0.1mM dATP containing 0.1 mM biotin-14-dATP in 500m Tris-HCl (pH 7.8)

  50 mM $MgCl_2$

  100 mM β- mercaptoethanol

  BSA (100 μg / ml).

- **Enzyme mix 10X**

  DNA polymerase 1 (0.5 units / μl BRL)

  DNase 1 (0.0075 U / μl)

  50 mM Tris (PH 7.5)

  5 mM Magnesium acetate

  1 mM β-mercaptoethanol

  0.1 mM Phenyl Methyl-Sulfonyl Fluoride (PMSF)

  50 % v/v Glycerol

  Nuclease free Bovine Serum Albumin (100 μg / ml).

- **Stop buffer**

  300 mM EDTA (pH 8.0).

- **BlueGene Non-radioactive Nucleic Acid Detection system**

  Streptavidin, Biotin (AP), NBT (nitroblue tetrazolium salt), BCIP (5-Bromo-4-chloro-3-indolyl phosphate (X-phosphate) from GIBCO/BRL.

  NBT dye mix: Add 33 μl of NBT solution to 7.5 ml of buffer C (given below).

- **Buffer A**

  Tris-HCl (0.1M, pH 7.5)

  NaCl (0.1 M)

  $MgCl_2$ (2mM)

- **Buffer B**

  Bovine serum Albumin (3 % w / v) in Buffer A.

- **Buffer C**

  Tris-HCl (0.1 M) pH 9.5

NaCl (0.1 M)

$MgCl_2$ (50 mM).

- **Column buffer**

  50 mM Tris (pH7.9)

  2mM EDTA (8.0)

  250 mM NaCl.

- **Sephadex G-5** soaked for 3 hours in water and suspended in coumn buffer.

- **20 X SSC solution**

  3 M NaCl

  0.3 M sodium citrate (trisodium citrate)

  Make upto 1 L with distilled water, adjust the pH to 7.0 and autoclave. Prepare 2X, 0.2 X and 0.16 X SSC buffers with 0.1% (w/v) SDS in each.

- **Triton X-100 (0.05 % v/v).**

- **Denatured single stranded salmon sperm DNA**: 100 g/ml (Sigma).

- **Denhardt's solution :**

  50 X (Add together 5g Ficoll, 5 g Poly Vinyl Pyrrolidone, 5 g BSA fraction V and bring upto 500 ml with distilled water. Filter through 0.45 μm membrane. Dispense in 25 ml portions and store at -20° C.

- **Prehybridization buffer (1.5 X)**

  117 g NaCl

  12.1 g Tris base

  0.7 g EDTA

  2 g Sarkosyl (N-lauroyl sarcosine)

  2 g sodium pyrophosphate

  Dissolve in distilled water and bring it upto 1.5 litres. Adjust pH to 7.9.

- **Stop buffer to terminate the dye detection reaction:**

  20 mM Tris (pH 7.5)

  5 mM EDTA

  Nitrocellulose paper and plastic hybridization bags or tubes.

  Ice, glass baking trays, plastic wrap, micropipettes etc.

Procedure

The procedure consists of: Preparation of the probe, labeling of probe, purification of probe (to remove unincorporated nucleotides), hybridization of probe and detection of the hybridized probe.

Preparation of the probe

Any DNA fragment used as an insert in gene transfer or which is complementary to the integrated gene present in the DNA on the filter can be used as a probe, since it readily hybridizes with the gene and indicates its presence. The probe DNA (fragment) is prepared by cutting it out with appropriate restriction enzymes and run on an agarose gel. The probe fragment can then be eluted from the gel and precipitated (refer to these procedures in earlier pages).

Labeling of probe

1. Prepare the labeling mixture by adding the following in a microcentrifuge tube: 5.0 μl dNTP 10 X mix. 1 μg DNA in TE buffer - to be labeled (denatured by heating in a boiling water bath for 10 minutes) Adjust the total volume to 45 μl. To this add 5 μl of the 10 X enzyme mix. Mix by tapping the tube gently.

2. Spin for three seconds at top speed to collect the contents together. Incubate at 16° C for 1 hour. In the meantime, prepare the column for purification of probe.

3. To prepare the column with Sephadex, take a 1 ml disposable syringe and remove its plunger. Push a small ball of glass wool into the syringe. Pack it well by pushing it with the plunger. Take a 15 ml Corex tube and place a microcentrifuge tube in it (after cutting off the cap). Position the syringe in the Corex tube such that the contents can be collected in the microcentrifuge tube. Pour the Sephadex in the syringe and centrifuge the assembly. The liquid passes down into the microcentrifuge tube and the Sephadex gets packed in the syringe. Add more Sephadex and centrifuge. Repeat this till the Sephadex column reaches 0.9 mark. Pour 10 μl of column buffer over the column and place a cean sterile microcentrifuge tube in the assembly.

4. Add 5 μl of stop buffer to the label mix after 1 hour and mix by gentle tapping.

5. Add 50 µl of column buffer to it and layer it over the column. Centrifuge for 5 minutes to collect the labeled purified probe. Store it at -20°C.

Hybridization of probe

1. Place the dry filter from a Southern blot in a plastic bag (a tray or a hybridization tube can be used) and add 15 ml of 1.5 X prehybridization buffer, 4 ml Denhardt's solution and 0.2 ml of Salmon sperm DNA.

2. Incubate the blot for 2-4 hours at 68° C by agitating the blot occasionally.

3. Dilute 50 µl of purified probe by adding 400 µl of distilled water to it in a microcentrifuge tube. Place this in a boiling water bath for 2 - 4 minutes and add to the bag. Incubate the filter in the probe mix at 68° C overnight.

4. Remove the filter and place it in a glass tray for washing. Incubate the filter with 250 ml of 2 X SSC with 0.1 % (w/v) SDS for 3 minutes and decant the wash solution. Repeat twice. Then repeat the wash twice each with 0.2 % SSC, and 0.16% SSC each containing 0.1 % (w/v) SDS. Finally, rinse the filter in 2 X SSC with 0.1 %(w/v) SDS at room temperature and blot it with Whatman 3 MM paper. Wrap in a plastic wrap or proceed with the detection of,the hybridized probe.

Detection of the hybridized probe

1. Incubate the filter in buffer B (taken in a tray) for 20 minutes at 42° C (a water bath may be used).

2. Prepare the antibody mixture as given below: Prepare 15 ml solution of Streptavidin in buffer A (2 µg / ml) i.e add 2 µl of stock (of 1 mg / ml) per 1.0 ml of Buffer A.

3. Incubate the filter in the antibody solution for 10 minutes by gentle agitation. Decant the solution (it can be reused) and wash the filter with buffer A for three minutes. Repeat twice.

4. Prepare 15 ml solution of Biotin (AP) solution in buffer A (1 µg / ml) i.e add Biotin solution for 10 minutes. Wash it thrice with buffer A.

5. Prepare the dye mixture (7.5 ml of dye solution added to 25 μl of BCIP solution). Incubate the filter in the dye solution in low light or dark for about 4 hours.

6. Wash the filter with the stop buffer to terminate the color development,

7. Photograph the filter and store in dark after baking it at 80° C in a vacuum oven for 1-2 minutes.

NORTHERN BLOTTING

The analysis of RNA is carried out by RNA blotting which is also called Northern hybridization (Alwine *et al.*, 1977). It basically consists of hybridization of RNA/DNA or sense RNA and antisense RNA hybridization. The denaturing agarose gel electrophoresis (Lehrach et al., 1977; Goldberg, 1980; Seed, 1982a) is utilized to separate RNA molecules according to their sizes which are then blotted onto nitrocellulose or nylon membranes. The membrane is then hybridized with a specific probe to detect the mRNA of interest through autoradiography or non-isotopic methods. Genomic DNA, cDNA, oligonucleotide sequences or antisense RNA can be used as the probe in Northern hybridization. RNAs are very mobile compared to the DNA and great care should be taken to maintain the purity It is very essential to minimize RNase activity. Therefore, gloves must be worn at all times. Whenever possible, plasticware should be autoclaved and all glassware and other containers should be treated with 0.1 % diethyl pyrocarbonate (DEPC) and autoclaved. Glassware should be baked overnight at 250° C. All apparatus should be cleaned with DEPC treated water and all reagents are to be prepared with DEPC treated water.

Electrophoresis of RNA and Northern transfer

The protocol for RNA isolation has been explained earlier. The RNA sample is then subjected to electrophoresis. Since RNA is single stranded and hence can have differing amounts of secondary structure it must be electrophoresed under denaturing conditions. Formaldehyde is generally used as a denaturant. Formaldehyde is a suspected carcinogen and hence must be used with caution and in a hood.

Reagents

- **10 X MOPS buffer**

0.2 M MOPS

50 mM sodium acetate

10 mM EDTA

adjust to pH 7.0 and autoclave.

- **Electrophoresis running buffer**

  (1 X MOPS buffer) Dilute the 10 X MOPS buffer.

- **Electrophoresis sample buffer**

  50 % formamide

  1 X MOPS buffer

  5.7 % formaldehyde

  5 % glycerol

- **20 X SSC**

  3 M NaCl

  0.3 M sodium citrate

  Formaldehyde

  Ethidium bromide (10 mg / ml)

Materials

Blotting kit, nitrocellulose or nylon membranes, 3 M Whatman filter paper and Saran wrap.

Procedure

1. Prepare a 1 % agarose solution with 0.6 g agarose and 45 ml DEPC treated distilled water and 5 ml 10 X MOPS buffer. Cool it to 50 - 60° C and add 3.5 ml formaldehyde to it. Add 2 μl EtBr and mix well. Prepare a gel mold, position the comb and pour the gel . Let it set for 30 min.

2. Mix together 5 μl of RNA sample solution and 20 μl of 1X RNA sample buffer and heat it at 65^0C for 15 min.

3. Remove the comb gently and set the gel in the electrophoresis tank. Fill the tank with the running buffer just to cover the gel. Flush the wells with the running buffer using a pipette and pre-run the gel for 10 min at 80 V (2-10 volts/cm).

4. Load the samples into the gel and electrophorese for about one hour at 80 V.

5. Photograph the gel using a transparent ruler on the UV transilluminator (use a saran wrap below the gel to prevent RNase contamination).

6. Rinse the gel with 10 X SSC by gentle shaking at 60 rpm for 20 min. Unlike the DNA gel, the RNA is already denatured due to the formaldehyde, so no denaturation is required.

7. Cut the nitrocellulose membrane or nylon membrane and one sheet of 3M whatman filter paper of the same size as the gel. Cut out another larger piece of 3 M Whatman filter paper. Handle the nitrocellulose paper with gloves and mark it on the upper left or right corner with a pencil.

8. Moisten the membrane and filter papers for 5 minutes in water and later with the 10 X SSC solution.

9. Take the blotting tray and place a large sponge. Pour the 10 X SSC solution (transfer solution) and saturate the sponge with it. Lay the bigger Whatman (pre wet) paper over the sponge. Alternatively, place the large Whatman filter paper on a flat elevated plate in such a way that the two ends hang into the transfer buffer and serve as wicks. Fill the blotting tray with the transfer buffer to the edge of the flat plate but do not allow the solution to come up over the plate !

10. Place the gel carefully on top of the filter starting from one side and with the well-side facing down. Run a pipette or glassrod gently over the gel to remove any air bubbles. Cover the rest of the area all around the gel with parafilm to prevent contact of the top paper towels with the wick.

11. Place the nitrocellulose membrane over the gel. The marked side should face the gel. Take care to prevent any air bubbles in between the gel and the membrane as they will interfere with the transfer.

12. Gently overlay the membrane with an exactly sized Whatman filter paper. Place a stack of paper towels on top of the gel assembly and put a light weight on top of the paper towels.

13. Leave over night for the transfer to be completed by capillary action. Remove the paper towels the membrane filter and use a ball point pen to

trace the position of the wells on the membrane before removing it from the gel.

14. Airdry the membrane and bake it at 80° C for about 2 hrs in a vacuum oven or overnight in a regular oven.

15. Store the membrane in a sealed cellophane bag at 4° C and prepare for the hybridization procedure.

RNA hybridization

The general procedures for prehybridization, hybridization, washing, detection and autoradiography are the same as described for Southern blotting except that all the solutions should be treated with DEPC.

DOT AND SLOT BLOTTING

The dot / slot blotting procedure (Kafatos et al., 1979) is much simpler and faster than the Southern and Northern blotting. The dot / slot blots are used in DNA / DNA, DNA / RNA and RNA / RNA hybridizations.In this method, the nucleic acid samples can be directly applied on the membrane filters without any electrophoresis. However, this test reveals only the presence of the target gene in the DNA sample but cannot reveal the sizes of specific bands of interest in the sample. But many samples give background signals that do not reflect true hybridization. Consequently, care should be tken to include appropriate controls.

Materials

- Vacuum blot apparatus
- Nitrocellulose or nylon membrane
- 3 M Whatman filter paper.

Reagents needed

- **20 X SSC solution**

  3 M NaCl

- 0.3 M sodium citrate

  Dissolve well, adjust pH to 7.0 and autoclave.

- **Denaturing solution**

  1.5 M NaCl

  0.5 M NaOH

  Dissolve in sterile distilled water.

- **Neutralizing buffer (for DNA dot blot)**

  1.5 M NaCl

  0.5 M Tris - HCl, pH 7.4

  1 mM EDTA

  Dissolve in sterile distilled water.

- **Neutralizing buffer (for RNA dot blot)**

  66 % (v/v) formamide

  21 % (v/v) formaldehyde

  13 % (v/v) 10 X MOPS buffer (refer to Northern blotting

Procedure

- Denature DNA by heating the samples to 95° C for 15 min and immediately, chill on ice. Spin briefly and add equal volume of 20 X SSC to it. Alternately, 0.2 volume of 2 M NaOH solution can be added to the sample and left for 15 min at room temperature, followed by the addition of 1 volume of neutralizing buffer (0.5 M Tris-HCl pH 7.5 and 1.5 M NaCl). Leave at room temperature for 15 min.

- Assemble the vacuum blot apparatus, cut out a piece of the nitrocellulose or nylon membrane and fix it in it. Apply slight vacuum only.

- Spot the samples in the provided slots under vacuum (use 5 µl of sample) on the membrane after prewetting it with 10 X SSC.

- Airdry the filter and place it on a 3M Whatman filter paper pre-saturated with denaturing solution. See that the DNA side is up. Incubate for 10 min.

- Then transfer the filter to a 3M Whatman filter paper pre-saturated with neutralizing solution for 5 min. Air dry the membrane for 20 min.

- Wrap the filter in a saran wrap and place DNA side down on a transilluminator for 4 -6 min to cause the DNA to bind tightly to the membrane. Alternatively, bake the filter in an oven at 80° C under vacuum for 2 hrs.

- The prehybridization, hybridization, washing and detection are the same as described for the Southern blotting.

Dot blotting of RNA

Heat the RNA samples at 65^0C for 10 min in five volumes of denaturing solution for RNA and quickly chill it on ice for 5 min. Spin down briefly . Add 1 volume of 20 X SSC and and spot it on the membrane filter. The subsequent processing is the same as described for Northern blotting.

WESTERN BLOTTING

Western blotting (Burnette, 1981 and Blake *et al.* 1984) is a procedure where antibodies react with proteins. It is used for immunodetection of specific antigens in a collection of proteins. This technique involves the separation of proteins based on their molecular weights with SDS-PAGE and their immobilization onto a nitrocellulose or nylon membrane. Then the specific protein (band) is identified by using a specific antibody raised by the specific antigen (in an animal e.g rabbit) to bind to it. This antigen-antibody complex is then detected by an enzyme linked assay consisting of a second antibody that is commercially available covalently linked to an enzyme.

The procedure of SDS-PAGE has been described in an earlier chapter. After the electrophoresis of the proteins is completed, blotting can be taken up.

Materials

Semi dry- transfer cell [e.g, Trans-Blot SD of Bio-Rad or any other make] and the instruction booklet supplied with it for reference.

Nitrocellulose or nylon membrane (Immobilon P)

3 M Whatman filter paper

Reagents

- **Blotting buffer**

  1.5 g Tris-base

  7.2 g glycine

0.5 g SDS

200 ml methanol

Bring up to 1 liter with distilled water and adjust pH to 8.3.

- **Phosphate buffered saline (PBS) tween**

  10 mM NaH2PO4 (1.38 g / l)

  150 mM NaCl (8.8 g / l)

  Adjust the pH to 7.2 and add 0.3 % tween - 20^0 before use.

- **AP buffer**

  100 mM Tris - base (12.1 g / l)

  100 mM NaCl (5.8 g / l)

  5 mM $MgCl_2$ (1 g / l)

- **Enzyme substrate stocks**

  17 mg BCIP (5-bromo-4-chloro-3-indolyl phosphate) dissolved in 500 μl dimethyl sulfoxide (DMSO). Dissolve in water if the BCIP is a sodium salt. 33 mg NBT (Nitroblue tetrazolium) dissolved in 1 ml 70 % DMSO.

  Store the stocks in the dark at room temperature.

  To prepare the enzyme substrate mix, add 50 μl of BCIP stock and 100 μl NBT stock to 10 ml of AP buffer.

  Wear gloves throughout! Methanol can be absorbed through the skin and also cause eye damage!

Procedure

1. Equilibriate the gel in blotting buffer for 15 min at room temperature. Soak the nitrocellulose membrane in the blotting buffer. Saturate four pieces of 3 M Whatman filter paper in the blotting buffer. Also soak the blotting sandwich.

2. Arrange 2 pieces of filter paper on a thin sponge resting on the clear plastic side (lower electrode) of the blotting apparatus. Then place the nitrocellulose membrane on them and cover with the gel followed by two pieces of the filter paper. Take care that no air bubbles form.

3. Close the sandwich. Place in blotting chamber with the clear plastic facing electrode (i.e the membrane side should face the electrode.
4. Blot overnight at 200 mA or 20 volts. Proceed with the development of the Western blot.
5. After the blotting, rinse the membrane in a tray with 20 ml PBS tween for 5 min. Repeat once.
6. Add 8 ml of the primary antibody to the PBS - Tween solution to a dilution of 1:250 and incubate the filter in it for 45 min at room temperature.Save the antibody solution. It can be used again.
7. Wash twice for 5 min each with 20 ml PBS-Tween and transfer to fresh PBS-Tween and add the secondary antibody at a dilution of 1:1000 and incubate at room temperature for 45 min.
8. Wash with PBS-Tween and later wash twice with the AP buffer (to remove the phosphate).
9. Add the enzyme substrate mix (developing solution). Stop the reaction when the coloured bands develop by rinsing with tap water.
10. Photograph the blot with a transparent ruler alongside or make a photocopy of the filter after wrapping it in saran wrap.
11. Calculate the molecular weights of the proteins that are recognized (linked) by the antibody by reading off from the MW graph of standard proteins.

As an example, run a β-galactosidase standard on a SDS-PAGE gel and probe it with antibodies to this protein. The antibody to β-galactosidase (can be obtained from 5 Prime to 3 Prime Inc) is diluted in 1:100 of TTBS/milk solution and used. The TTBS / milk can be prepared by adding 1 g of non-fat, dry milk in 10 ml of TTBS. The TTBS or Tween Tris Buffered Saline can be prepared by adding 0.5 ml of Tween 20 to 1 litre of TBS (20 mM Tris and 500 mM NaCl in 1 litre of distilled water and pH adjusted to 7.5). Protein A-peroxidase can be used as the secondary antibody by diluting it 1:500 into TTBS /milk. To detect the enzyme by color development, use 3,3'-diaminobenzidine solution. This is prepared by adding 10 μl of 30 % (v/v) H_2O_2 to10 ml of the 3, 3'-diaminobenzidine-Tris solution (0.5 mg 3, 3'-diaminobenzidine in 1 ml of 0.05 M Tris and adjusted to pH 7.6).

3, 3'-diaminobenzidine is a carcinogen ! It must be used with caution ! Use gloves !!!

The color develops within 5 to 10 min. Rinse with distilled water to stop the reaction.

PCR AMPLIFICATION OF DNA

The polymerase chain reaction (PCR) is a powerful method used in molecular biology (Saiki et al., 1985, 1988; Mullis et al., 1986; Mullis and Faloona, 1987; Chien et al., 1976). By employing this method, extremely small quantities of DNA (nanograms) can be rapidly amplified and analyzed. PCR is a relatively new technique that has enabled the molecular cloning and analysis of DNA and RNA and in recent years it has been applied in the diagnosis of genetic disorders, the detection of specific nucleic acid sequences from pathogenic organisms in clinical samples (e.g., HIV) and the genetic identification of forensic samples. It is also used for preparing specific sequences for making probes, for the generation of cDNA libraries, for the generation of large amounts of DNA for sequencing and for the analysis of specific mutations. PCR is also used for assessments of genetic variability of germplasm, and can provide molecular markers for linkage studies. PCR can be used for a quick verification of successful integration of foreign DNA. Since extremely small quantities of DNA are used for PCR (as compared to restriction analysis and / or Southern blotting), the preparations do not need to be as "clean" (i.e., cellular contaminants that would disturb enzyme activity are diluted out for PCR).

In the analysis of DNA, PCR can lead to enormous amplification of the target gene. PCR is carried out in a thermocycler (Fig-13). PCR is specifically used to amplify the segment of DNA that lies between the two regions of known sequences. The most important step is to choose highly specific primers, which are complimentary to these known sequences of the target sequence between them. The template DNA is first denatured by heating in the presence of the two primers and the dNTPs. The mixture is then cooled to facilitate the annealing of the primers to the complementary regions. The primers are then extended with the DNA polymerase to yield the newly synthesized copies of template DNA.The products of one cycle of amplification serve as templates for the next cycle and the amount of the DNA product doubles after each cycle. The cycle of denaturation, annealing and DNA synthesis is repeated many times (about 35 cycles!).The specific type of polymerase used for the automatic replication cycles is the thermostable AmpliTaq polymerase from the thermophilic bacterium *Thermus aquaticus* which can withstand the high temperatures of the denaturation cycles.

The PCR can amplify any contaminating DNA that comes in contact with the samples which can be a source of errors. Therefore precautions need to be taken in labeling the reaction tubes accurately, using new pipet tips at every step and by wearing clean gloves at all times while preparing the samples. The following procedure describes the amplification of the β-glucuronidase (GUS) gene to confirm its presence in a putative transgenic plant.

Requirements

- Thermocycler
- **DNA sample**

  Genomic DNA isolated from the putative transgenic plant (transformed with a construct including the GUS gene, which is the template DNA)
- **Positive control**

  DNA of the plasmid DNA construct carrying the GUS gene fragment.
- **Negative control 1**

  Genomic DNA isolated from a non-transgenic plant.
- **Negative control 2**

  Sample lacking DNA.
- **Target gene (GUS)**

  The gene present in the construct that was transferred into the plant cells.
- **Primers for GUS gene :**

  Primer I: CTTTAACTATGCCGGAATCCATCG

  Primer II: TAACCTTCACCCGGTTGCCAGAGG

  These primers amplify a 250 bp fragment within the GUS gene.
- d NTPs(a mixture of all the four deoxynucleotides) (1.25 mM).
- Taq DNA polymerase (from Perkin Elmer Amplitaq. Recombinant Taq DNA polymerase)

- Pure mineral oil
- Amplification buffer (10 X): 500 mM KCl, 15 mM Mg Cl_2, 100 mM Tris, pH = 8.0.

Precautions

This procedure needs to be done very carefully without risking any contamination by foreign DNA. Use gloves and change them frequently. Sterile pipet tips and tubes to be used. Use fresh reagents only.

Procedure

1. Prepare a master mix by mixing the designated amounts of each reagent (sufficient for all the samples) in a 0.5 ml microcentrifuge tube. Multiply the amounts according to the number of assays (use a 1.5 ml microcentrifuge tube if the quantity exceeds 400 µl).

   - **Master mix**: (for each 20 µl sample)

     2 µl 10X PCR buffer (with $MgCl_2$ concentration, 1 mM)

     1 µl primer I

     1 µl primer II

     1.6 µl, 2.5 mM dNTPs

     0.1 µl diluted Taq polymerase (5 U / µl)

     14 µl sterile water

2. Distribute the 19.7 µl of the master mix in each of the sample tubes (sterile microcentrifuge tubes). Add the samples (The genomic DNA being analysed, positive control and negative control) and add 25 µl of light mineral oil to the top of each. The mineral oil acts as a barrier to prevent evaporation when the tubes are heated.

3. Close the tubes tightly. Spin tubes for 5 seconds in a microcentrifuge (use carrier tubes).

4. Place all the tubes on ice and carry them to the thermocycler (PCR machine). Put the tubes firmly in the wells and close the lid.

5. Turn on the machine. If the following program is not in the machine, create a new file so that following step 1, there will be 35 cycles with

the specifications given below: Step 1 : **94° C for 5 min** (to inactivate proteases in the extracts) one time only.

Cycles : **94^0C for 15 seconds (denature)**

: **60^0C for 15 seconds (anneal)**

72^0C for 60 seconds (polymerize...extension of primers)

Hold at 72^0C for 5 min and hold at 4^0C until the sample is removed.

6. Wait for the completion of the PCR reactions which might take several hours and remove the samples. Insert a long tipped pipet into the bottom of each tube and withdraw the sample leaving behind the oil phase on the top. Wipe the tips to clear it of the oil and transfer each sample to a fresh sterile tube.

7. Check the DNA samples by electrophoresis (follow the procedure described earlier). Prepare a 1 % agarose gel in 0.5 X TBE (with ethidium bromide). Add 5 μl of loading dye to each of the samples and load 15 μl of sample into each well. Run the electrophoresis at 100 V until the dark blue dye front moves 3/4 down the gel. Photograph the gel on the UV transilluminator (Fig - 27).

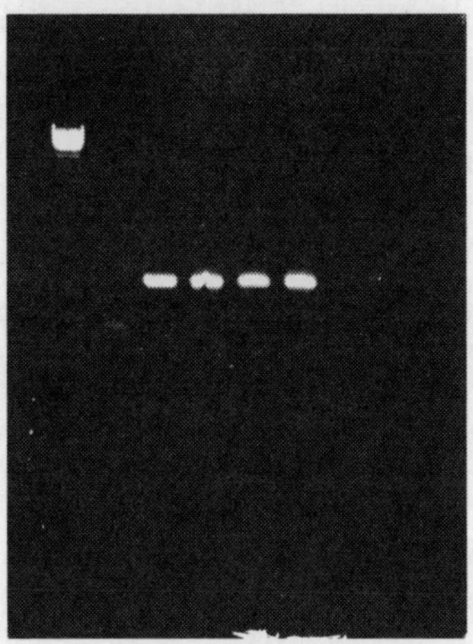

Fig. 26. Gel with PCR amplified DNA.

8. A Southern blot can be carried out and later the filter hybridized with the specific probe for further confirmation of the DNA (it can be compared with the positive and negative controls) or the DNA can be eluted from the agarose gel (from the specific fragment) purified and the concentration measured.

9. *Trouble-shooting*: If no amplification is obtained, the following should be tried : $MgCl_2$ concentration, (1-5 mM) ; dNTP concentration, (0.2-2 mM); less or more template DNA ; Change length of steps in cycle, (15 sec-2 min); change annealing temperature (35° -65°C).

CDNA CLONING BY RT - PCR

RT - PCR can be used for cDNA cloning. This method (Frohman *et al.* 1988., Gilliland *et al.* 1990) is much simpler and inexpensive compared to the other methods. Total RNA or mRNA is first reverse transcribed into cDNA that is then amplified by PCR. Specific primers are needed in relation to the arrangement of the amino acids in the protein. The primers will anneal to the first cDNA strand that is synthesized from mRNA by reverse transcription. However, the cDNA obtained is usually only a part of the actual whole length message depending on the distance between the two primers. Sometimes the primers may anneal to non-specific sequences producing artifacts. Therefore the PCR products need to be verified by dot blot or Southern blot hybridization.

Requirements

The same as described for the PCR method. In addition, the enzyme reverse transcriptase (about 20 - 40 units per 20 ng RNA) is needed to synthesize cDNA from the RNA.The buffer is the same.

Procedure

The first step involves the reverse transcriptase generation of cDNA followed by the next step of PCR amplification of the cDNA strand.

Reverse transcriptase generation of cDNA:

1. Take a sterile microcentrifuge tube and add together the RNA, primers, buffer and water.

2. Heat the contents to 65 ^{0}C for 5 min.
3. Cool on ice for 5 min. Add the dNTPs and reverse transcriptase (to a final volume of 20 µl).
4. Incubate at 42^{0}C for 40 min.
5. Add 80 µl of 10 mM Tris - HCl, 1 mM EDTA (pH 8.0) to stop reaction.
6. Remove 2 µl of the synthesized cDNA and amplify it using the PCR (follow the method previously described).

One step RT - PCR procedure (direct addition of reverse transcriptase and polymerase)

1. Add together the following in a sterile microcentrifuge tube :

   2 µl RNA (5 ng / ml)

   2.5 µl of primer I

   2.5 µl of primer II

   2.5 µl amplification buffer (with $MgCl_2$ 1-5 mM)

   34.75 µl water
2. Heat to 65° C for 5 min, cool on ice for 3 min.
3. Add the following to each tube :

   2.5 µl dNTPs

   2 µl AMV reverse transcriptase (10-20 U/µl, Dupont NEE-127)

   1.25 µl diluted Taq polymerase
4. Place in the thermocycler and set to 30 - 50 cycles as follows :

   42^0 C for 40 min (one time only) cDNA synthesis

   94^0 C for 1 min

   55^0 C for 1 min

   72^0 C for 1 min
5. Check the product by running on a 1 % agarose gel.

Note: The Taq polymerase can withstand the extreme temperatures because it is isolated from a thermal bacterium. The reverse transcriptase activity will be "killed" by the 94° C treatment.

RAPD ANALYSIS

Randomly amplified polymorphic DNA analysis (RAPD) is used for genome mapping, gene tagging, and studies of phylogenetic relationships (Williams *et al.* 1990). In this analysis, DNA fragments are synthesized in a PCR reaction mixture of genomic DNA and a randomly chosen primer that has binding sites on the complementary strands of the DNA within approximately 3 Kb. It is not essential for the template DNA to be used for the PCR to be of a high molecular weight. Therefore simpler methods of DNA isolation can be employed for the RAPD analysis.

Requirements

- DNA sample (20 ng / µl)
- **PCR buffer**

  10 mM Tris-HCl . pH 8.8.

  50 mM KCl,

  1.5 mM $MgCl_2$

  0.1 % Triton X - 100
- **dNTPs**

  0.1 mM each of dATP, dCTP, dGTP and dTTP
- **Primer**

  Random primer (this has to be decided after preliminary trials. Individual oligonucleotide and pairs of oligonucleotides are commercially available.)
- Taq DNA polymerase

Procedure

1. Prepare a master mix (explained in the PCR protocol) by adding the following:

   2 µl PCR buffer (with $MgCl_2$ 1-5 mM)

2 μl random primer

2 μl dNTPs

0.3 μl Taq polymerase (5 U / μl)

14 μl sterile water

2. Add 25 μl mineral oil and spin for 5 sec.
3. Run the PCR program at the specifications mentioned in the PCR protocol.

   (These specifications can be modified by changing the time for each step since this can vary with the type of primer . Follow the trouble shooting tips provided with the PCR protocol.)
4. Electrophorese the sample on a 1% agarose gel and check the DNA.

GENE CLONING AND RECOMBINANT DNA TECHNIQUES

The study of gene function coupled with the discovery of restriction endonucleases have opened up immense opportunities for scientists to develop methods of studying the molecular basis of genetics, which have increasingly centered on recombinant DNA techniques. The recombinant DNA technology which includes almost any technique for manipulating DNA or RNA has produced a wealth of knowledge in all areas of biology and has paved the way for the establishment of biotechnology, devoted to the production of drugs, vaccines and several other important useful products.

The most important and fundamental aspect of the recombinant DNA technology is gene cloning. Gene cloning includes:

1. Isolation of a DNA sequence or gene of interest.
2. The splicing of this DNA fragment into a cloning vector or vehicle such as a bacteriophage or plasmid.
3. Introduction of this recombinant plasmid into a host cell (usually bacterial cells) where it is replicated many times in an amplification process known as cloning.

4. The gene or insert can be isolated from the clones for further use. The cutting of the DNA is carried out by restriction endonucleases, followed by ligation with DNA ligases (Fig - 12).

The whole of the genomic DNA isolated from a plant can be digested with a suitable restriction enzyme and the DNA fragments ligated into a collection of vectors which are also cut by the same restriction enzyme. This collection is referred to as a genomic library. A cDNA library can also be prepared by digesting the cDNA and cloning in vectors. The polymerase chain reaction (PCR) used for amplification of specific DNA sequences is also used for gene cloning.

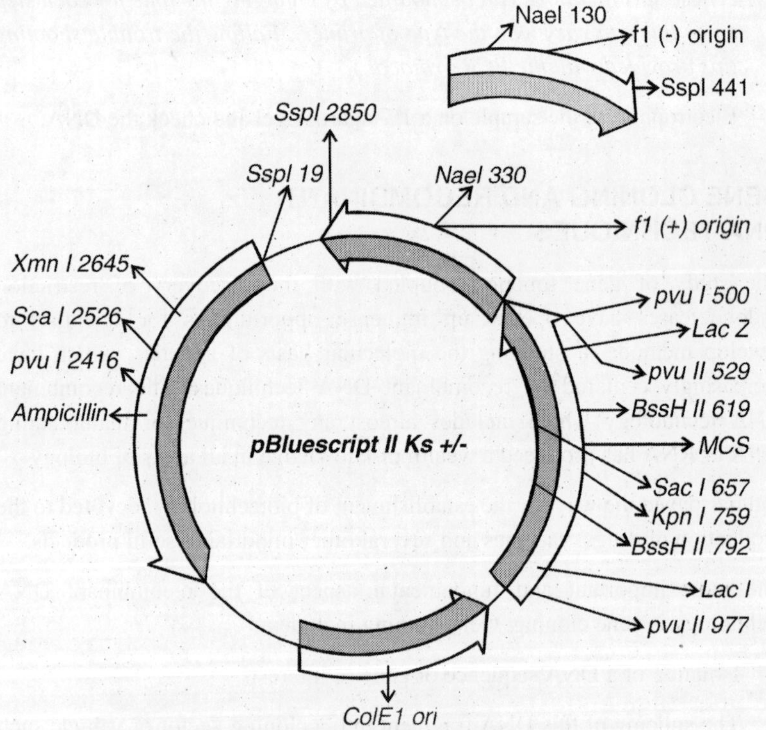

Fig . 27: Restriction map of p Bluescript II KS+ /-

The experiment described here deals with an *E. coli* strain carrying the p BlueScript II KS +/- plasmid (phagemid). The BlueScript phagemid vector (Fig. 28) has an ampicillin resistance gene, a multiple cloning site and the *lac* Z gene. The *lac* Z gene can be induced by isopropyl thiogalactoside

(IPTG) to express β-galactosidase. The enzyme has the ability to hydrolyse the colourless 5-bromo-4-chloro-3 indolyl-β-D-galactoside (X-gal) and produce a blue colour. The multiple cloning site is in the *lac Z* gene region. Insertion of a foreign DNA fragment in the multiple cloning site leads to the loss of β - galactosidase activity of the host strains (*E. coli*). After transformation, selection of bacterial transformants is done on the basis of the colour of the colonies produced on X-gal plates. If colonies appear white, it indicates a foreign DNA insertion (i.e they carry a recombinant plasmid) and if they appear blue, then it means that there is no insert in the vector.

The gene cloning methods are explained in the following steps:

- Isolation of the plasmid DNA (p Blue script II KS +) from an *Escherichia coli* strain (host).
- Digestion of the plasmid DNA and the λ – DNA (to generate a fragment to be used as an insert) with the restriction endonuclease Eco RI.
- Ligation of plasmid DNA and insert DNA (the fragment created by Eco R1 from the λ – DNA) to produce a recombinant plasmid.
- Preparation of competent cells of the *E. coli* host.
- Transformation of *E. coli* cells with the recombinant plasmid.
- Selection of transformants.
- Analysis of the recombinant plasmid by restriction digestion.

Isolation of plasmid DNA:
(p Bluescript II KS + from the specific *E. coli* strain)

Procedure

Isolate the plasmid DNA from the bacterial host cells according to the method described earlier in this chapter.

Digestion of plasmid DNA and insert DNA with the restriction endonuclease Eco R1

Procedure

1. Digest both the DNA with Eco R1 by following the protocol described ealier in this chapter.
2. Check by electrophoresis of the digests on 1 % agarose gels.

3. Study the DNA fragments (bands) under U.V light.
4. The uncut plasmid DNA being circular, migrates faster than the others. The digested plasmid DNA (it has only a single Eco RI site) will be linearized and appears as a band above the uncut DNA. The size of the linearized plasmid can be ascertained with the help of the λ - Hind III marker run alongside. Similarly, the Eco R1 fragment from the λ– DNA (the insert) can be checked by comparing it with the DNA size marker ladder.
5. Determine the volume of insert DNA to obtain 500ng (0.5 μg) and the volume of the vector DNA to obtain 100 ng. A 5:1 or 10:1 molar excess of the insert fragment is advisable.

Ligation of plasmid and insert DNA to produce a recombinant plasmid

After isolation and digestion of the plasmid (vector) DNA with the restriction enzyme, the cleaved DNA is joined to the insert DNA *in vitro.* This process is called ligation and requires the enzyme DNA ligase.

The plasmid DNA (which was cleaved and linearized) can be joined with the restriction fragment generated from the digestion of the λ – DNA with Eco R1. The recombinant vectors thus formed will be circular but bigger than the non-recombinant vectors. They will contain the gene for ampicillin resistance but the lac Z gene will be disrupted by the insert. So these will not form blue colonies on X-gal + IPTG medium.

For perfect cloning, the ends of the cleaved vector should ligate to the insert fragment. However, ligation can join any two compatible fragments i.e in particular, recircularization of the vector (without the insert) can happen. Therefore an optional process called dephosphorylation of the digested vector DNA can be carried out where the phosphate group from one end of the vector DNA is removed by an enzyme, alkaline phosphatase.This prevents re-circularization. The insert fragments can however ligate with the plasmid even if its one end is dephosphorylated.

Dephosphorylation of the digested vector (plasmid) DNA

(Optional step)

1. Prepare the dephosphorylation mixture by mixing together the following:
   - Eco R1 digested plasmid DNA (take appropriate amount as suggested above).
   - 5 μl of 10 X calf intestine alkaline phosphatase(CIAP) buffer (supplied with enzyme).
   - units CIAP/ pmole of ends.
   - Add sterile dH_2O to make up 50 μl.

   The CIAP and 10 X CIAP buffer should be kept at 4^0 C and the reaction should be set up at 0^0C.

   Calculate the ends as follows: Use the formula

$$\text{p mol end} = \frac{\text{amount of DNA}}{\text{base pairs X 660 X 1000}} \text{X 2}$$

   (The base pairs is deduced from the size of the fragment.)

2. Incubate for 60 min at 37^0C. Stop the reaction by adding 1 μl of 0.5 M EDTA and heating to 65^0C for 20 min.
3. Extract the DNA with an equal volume of phenol : chloroform(1:1). Extract the aqueous phase again with an equal amount of chloroform : isoamyl alcohol(24:1).
4. Precipitate the DNA by adding two volumes of ethanol to the aqueous phase. Collect the precipitate by centrifugation for 20 min (at 4^0C) at 10,000 rpm and suspend it in 50 μl of sterile dH2O.

Ligation of plasmid and insert DNA

Reagents

- T4 DNA ligase enzyme

- 10 X ligase buffer (which includes ATP, and is supplied with enzyme) Sterile dH_2O

Procedure

1. Combine the purified cut plasmid DNA with the insert DNA in the proportions suggested above and co-precipitate the DNA by adding 1/10 volume of 3 M sodium acetate and 2 volumes of ethanol. Leave on ice for 30 min to overnight.

2. Centrifuge to pellet the mixed DNA. Be sure to use sterile procedures since the ligation mix will be used to transform bacterial cells. Resuspend the pellet in 7 μl of sterile water. Add 2 μl of the ligation buffer and 1 μl of T4 DNA ligase (1 unit) and incubate overnight at 16^0C. Store at 4^0C until use. Include a *no-ligase* control.

3. The recombinant molecules can be tested before transformation into the bacterial host cells. This can be done by electrophoresis along with the Eco R1 cut vector and the *no- ligase* control. The recombinant DNA forms a band well above the Eco R1 digested vector's band.

Preparation of competent cells of the *E. coli* host

Competent cells are those that have been treated in a special way with chemical ($CaCl_2$, polyethylene glycol, dimethyl sulfoxide and magnesium ions) or electrical agents (high voltage electrical pulses) so as to enable them to take up DNA (Mandel and Higa, 1970;Cohen *et al.*, 1972; Oishi and Cosloy, 1972; Hanahan, 1983). The treatment of bacterial cells with $CaCl_2$ is described here. This treatment induces a transient state of "competence" in them, enabling them to take up the insert DNA. The addition of dimethyl sulphoxide (DMSO) in the final step enhances the rate of transformation as DMSO acts as a carrier molecule for an introduced foreign DNA.

Reagents

- LB medium (pH = 7.0) (given in earlier pages)
- **1 M $CaCl_2$**
  147g of $CaCl_2$ made upto 1 liter with distilled water.
- **0.1 M $CaCl_2$**
  Make up 100ml of 1 M $CaCl_2$ to 1 liter with distilled water.
- Dimethyl sulphoxide reagent.

Procedure

1. Grow a fresh plate (LB medium) of E. coli from a frozen culture. *(Some strains of E. coli are capable of higher competence than others. The DH 5 alpha strain is a good one).*

2. Transfer a single colony of the E. coli to 100 ml LB medium (liquid) and grow the bacteria on a rotatory shaker overnight at 200 rpm at 37 °C.

3. Check the culture and monitor till it reaches an OD 600 of 0.3-0.4 units.

4. Transfer the medium containing the bacteria to sterile centrifuge tubes and cool on ice for 15 min.

5. Centrifuge at 5000 rpm for 5 min at 4° C. Discard the supernatant and add 10 ml ice-cold 0.1M $CaCl_2$, gently resuspend the cells and store on ice for10 min.

6. Centrifuge at 4000 rpm for10 min at 4° C to recover the cells (pellet). Resuspend the pellet gently in 2 ml of ice-cold 0.1 M $CaCl_2$ per 50 ml of original culture.

7. Add 140 µl DMSO/4 ml of resuspended cells, mix gently and store on ice for 15 min. Add another 140 µl DMSO and dispense 100 µl aliquots in microfuge tubes. They can be used immediately for transformation or stored at -70° C in liquid nitrogen.

Transformation of *E. coli* cells

Transformation of the *E. coli* cells with the recombinant plasmid (with antibiotic marker) can be carried out by incubating the competent cells with the recombinant plasmid and by administering a heat shock. The DMSO acts as a carrier molecule for the transfer of the vector into *E. coli* cells.

Procedure

1. Take out four aliquots (100 µl) of competent *E.coli* cells from the freezer and thaw on ice. To three microfuge tubes, add the following as indicated:

   - Add 5 µl of ligation mixture to one aliquot
   - Add 1 µl of 1 µg / µl p Bluescript II KS + DNA to the second aliquot (positive control).

- Do not add any DNA to the third aliquot and to the fourth aliquot, add
- 5 μl of the *no-ligase* sample (negative controls).

2. Mix each sample gently by pipetting up and down with the pipet used to add the DNA.
3. Incubate all the tubes on ice for 30-60 min.
4. Incubate in a 42° C water bath for 60 sec (heat shock to increase the permeability of the bacterial membrane to the DNA).
5. Place on ice for 2 min.
6. Add 400 ml of sterile LB medium (prewarmed to 37° C) to each of the microfuge tubes and incubate at 37° C for 60 min (in a water bath). This helps in the recovery of cells and allows time for the expression of the antibiotic resistance gene.
7. Spread the contents of each tube on X-gal + amp plates for the selection of transformants (see next section).

Selection of transformants

After the transformation of *E. coli* cells with the plasmid DNA, initial incubation at 37° C in a rich medium permits cell growth before plating onto a selective medium. The selection of transformants is done by the screening of blue and white bacterial colonies. The bacterial colonies expressing the *lac* gene (blue colonies) grow poorly compared to those that don't express it (white colonies). However no blue colonies will appear if de-phosphorylaion of vector DNA was done.

Isopropyl -1-D-thio-β-galactoside (IPTG) is an inducer for the *lac* gene and allows for the expression of the lac gene to produce β - galactosidase. This enzyme cleaves 5-bromo-4-chloro-3-indolyl-β-D-galactoside (X-Gal) which is colorless, into galactose and the deep blue compound 5-bromo-4-chloro-indigo. This results in the blue colored colonies.

Reagents

- **Antibiotic**

  Prepare 10 μg/ml solution in dH_2O (ampicillin).

- **Isopropyl -1-D-thio-β-galactoside (IPTG)**

  Prepare 2 ml of IPTG solution by dissolving 80 mg in 2 ml of dH_2O. Filter, sterilize it and store in 0.25 ml batches in the freezer.

- **5 - bromo-4-chloro-3 indolyl-β-D-galactoside (X-Gal)**

  Prepare 1 ml of the solution by adding 40 mg X-Gal/ml of DMF (dimethyl formamide). Filter, sterilize and store in a glass tube in the freezer. Protect from light and wear gloves when preparing the solution. Make it fresh on the same day.

Procedure

1. Prepare X- Gal plates as indicated:

   - Prepare 250 ml portions of LB agar medium in 500 ml flasks and autoclave.

   - Cool to 50° C and add to each flask aseptically the following filter sterilized solutions:

     50 µl antibiotic solutio

     0.25 µl IPTG solution

     0.25 µl X- Gal solution.

   - Mix well and pour the X- Gal agar medium into sterile petridishes. Wrap the

   - X- Gal plates with aluminium foil and store the cooled plates at 4° C.

2. Shake the contents of the microfuge tubes (containing the transformed cells) which were incubated at 37° C for 60 min for recovery and expression of the antibiotic resistance gene.

3. Plate the transformed cell suspension onto the X-Gal plates with a sterile, flamed spreader (20-100 µl of suspension / plate).

4. Allow the agar to absorb the liquid (leave the plates on the lab bench for 15 min), invert the plates and incubate at 37^0C for 15 - 18 hours.

5. Blue and white colonies appear on the plates. Store the plates if necessary after sealing with parafilm.

6. Analyse the transformation as follows:

- Count the number of white (transformed) and blue (non-transformed) colonies.

- Calculate the transformation efficiency and the percent insertional inactivation from the ligation mixtures as follows:

$$\text{Transformation efficiency} = \frac{\text{Total number of colonies on the positive control}}{\mu\text{ g of plasmid plated}}$$

$$\text{Insertional inavtivation} = \frac{\text{Number of recombinants (white)}}{\text{Total (blue + white)}}$$

- Pick up the white colonies off the agar plates using a heat sterilized wire loop or sterile toothpick and inoculate into liquid and agar LB media for further work.

Analysis and confirmation of the recombinant DNA

Confirmation of the recombinant plasmid can be carried out by first isolating it and later the specific insert.

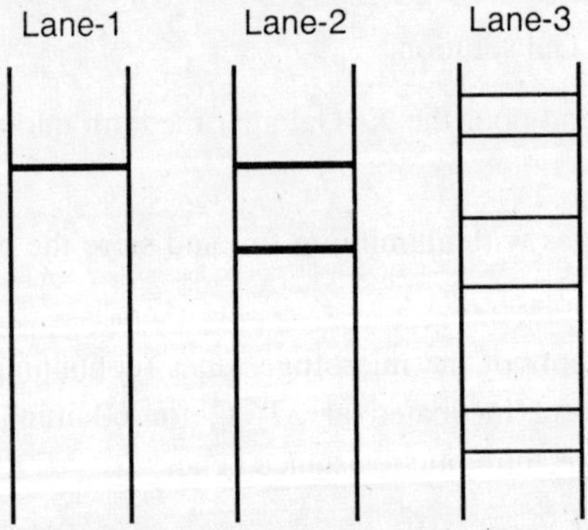

Lane-1 Lane-2 Lane-3

Fig 28: Analysis of DNA bands on a gel (diagrammatic presentation).

Lane 1: Eco RI cut non-recombinant plasmid = 1 band

Lane 2: Eco RI cut recombinant plasmid = 2 bands. The upper band concides with the band of the former. Therefore, the loser band is od the insert.

Lane 3: λ-Hind III maeker – to identify the size of the insert

Isolation of recombinant plasmid from the clones

Isolate the recombinant plasmid DNA from the clones by following the plasmid isolation procedure described earlier. Run the plasmid and the uncut non-recombinant plasmid DNA side by side on the agarose gel. If the band of the recombinant vector appears above that of the level of the uncut vector, it is likely to be a recombinant clone.

Isolation of the insert

The size of the insert needs to be ascertained. For this purpose, digestion of the recombinant plasmid with Eco RI enzyme and electrophoresis is necessary.

Follow the procedure described earlier.

- View the gel under UV light.
- Photograph the gel and analyze it as follows:

The Eco RI digest of recombinant vector should produce two bands (Fig - 28). Of these two, one should coincide with the band produced by the Eco RI cut vector DNA. The other band is the insert fragment. The molecular weight of the insert can be calculated by using the Lambda-Hind III digest marker. The insert DNA can be electro eluted for further characterization.

REFERENCES

1. Alwine. J. C., Kemp. D. J. and Stark. G. R. (1977) Method for detection of specific RNAs in agarose gels by transfer to diazobenzyloxymethyl paper and hybridization with DNA probes.Proc. Natl. Acad. Sci. 74 : 5350.
2. Arrand. J. E. (1985) In : Hames. B. D. H. and Higgins. S. J. eds. Nucleic acid hybridization : A practical approach. IRL Press. Oxford. 17.
3. Becker. J. M. Caldwell. G. A. Zachgo. E. A. (1996) Biotechnology. A lab course. 2nd edition. Academic Press. Ltd, London.
4. Birnboim. H. C. and Doly. J. (1979) A rapid alkaline extraction procedure for screening recombinant plasmid DNA. Nucleic Acids Res. 7: 1513 - 1518.
5. Blake. M. S., Johnston. K. H., Russel-Jones. G. J. and Gotschlich. E.C. (1984) A rapid sensitive method for detection of alkaline phosphatase-

conjugated anti-antibody on Western blots. Anal. Biochem. 136. 175 - 179.

6. Bonner, W. (1957) Autoradiograms.35S and 32P. Methods in Enzymology.152 : 55- 61.
7. Burnette. W. N. (1981) Western blotting : Electrophoretic transfer of proteins from sodium-dodecyl sulphate- polyacrylamide gels to unmodified nitrocellulose and radiographic detection with antibody and radioiodinated protein. Amer. Analyt. Biochem. 112, 195 - 199.
8. Chien. A., Edgar. D. B. and Trela. J. M. (1976) Deoxyribonucleic acid polymerase from the extreme thermophile Thermus aquaticus. J. Bacteriol. 127 : 1550.
9. Cohen, S. N., Chang, A. C. Y and Hsu, L. (1972) Non-chromosomal antibiotic resistance in bacteria : Genetic transformation of Escherichia coli by R-factor DNA. Proc. Natl. Acad. Sci. USA. 69 : 2110.
10. Dellaporta. S. L., Wood. J. and Hicks. J. (1983) A plant DNA minipreparation.: Version II. Plant Mol. Biol. Rep. 1 : 19 - 21.
11. Feinberg, A. P. and Vogelstein, B. (1983) A technique for radiolabelling DNA restriction endonuclease fragments to high specific activity.Anal. Biochem. 132 : 6.
12. Feinberg, A. P. and Vogelstein, B. (1984) Addendum. A technique for radiolabelling DNA restriction endonuclease fragments to high specific activity. Anal. Biochem. 137 : 266.
13. Frohman, M., Dush, M and Martin, G. (1988) Rapid production of full length cDNAs from rare transcripts : amplification using a single gene-specific oligonucleotide primer. Proc. Natl. Acad. Sci. USA 85 : 8998 - 9002.
14. Gilliland, G., Perrin, S., Blanchard, K and Bunn, H. F. (1990) Analysis of cytokine mRNA and DNA detection and quantitation by competitive polymerase chain reaction. Proc. Natl. Acad. Sci. 87 : 2725 - 2729.
15. Glisin, V., Crkvenjakov, R. and Byus, C. (1974) Ribonucleic acid isolated by cesium chloride centrifugation. Biochemistry 13 : 2633.
16. Goldberg, D. A. (1980) Isolation and partial characterization of the Drosophila alcohol dehydrogenase gene. Proc. Natl. Acad. Sci. USA. 77 : 5794.

17. Hanahan, D. (1983) Studies on transformation of Escherichia coli with plasmids. J. Mol. Biol. 166 : 557.
18. Kafatos, F. C., Jones, C. W and Efstratiadis, A. (1979) Determination of nucleic acid sequence homologies and relative concentrations by a dot hybridization procedure. Nucleic Acids Res. 7 : 1541.
19. Kricka, L. J. (1992) Non-isotopic DNA probe techniques. Academic Press. San Diego, California.
20. Langer, P. R., Waldrop, A. A. and Ward, D. C. (1981) Enzymatic synthesis of Biotin labelled polynucleotides : Novel nucleic acid affinity probes. Proc. Natl. Acad. Sci. USA. 78 : 6633 - 6637.
21. Leary, J. J., Brigati, D. J. and Ward, D. C. (1983) Rapid and sensitive calorimetric method for visualizing Biotin labelled DNA probes hybridized to DNA or RNA immobilized on nitrocellulose.: Bioblots. Proc. Natl. Acad. Sci. 80 : 4045 - 4049.
22. Lehrach, H., Diamond, D., Wozney, J. M. and Boedtker, H. (1977) RNA molecular weight determinations by gel electrophoresis under denaturing conditions, a critical re-examination. Biochemistry. 16 : 4743.
23. Mandel, M. and Higa, A. (1970) Calcium - dependent bacteriophage DNA infection. J. Mol. Biol. 53 : 159.
24. Mullis, K. B and Faloona, F. A. (1987) Specific synthesis of DNA in vitro via a polymerase - catalyzed chain reaction. Methods Enzymol. 155 : 335.
25. Mullis, K. B. Faloona, F. A., Scharf, R., Saiki, R., Horn, G. and Erlich, H. (1986) Specific enzymatic amplification of DNA in vitro. The polymerase chain reaction. Cold Spring Harbor Symp. Quant. Biol. 51 : 263.
26. Murray and Thompson (1980) Nucl. Acid Res. 8 : 4321.
27. Oishi, M. and Cosloy, S. D. (1972) The genetic and biochemical basis of the transformability of Escherichia coli K 12. Biochem. Biophys. Res. Commun. 49 :1568.
28. Parker, R. C. and Seed, B. (1980) Two dimensional agarose gel electrophoresis "Sea Plaque" agarose dimension. Meth. Enzymol. 65 : 358.
29. Prober, J. M. et al (1987) Science. 238 : 336.

30. Saiki, R. K., Scharf, S., Faloona, F., Mullis, K. B., Horn, G. T., Erlich, H. A. and Arnheim, N. (1985) Enzymatic amplification of β-globin genomic sequences and restriction site analysis for diagnosis of sickle cell anemia. Science. 230 : 1350.

31. Saiki, R. K., Gelfand, D. H., Stoffel, S., Scharf, S., Higuchi, R., Horn, G. T., Mullis, K. B. and Erlich, H. A. (1988) Primer-directed enzymatic amplification of DNA with a thermostable DNA polymerase. Science. 239 : 487.

32. Sambrook, J., Fritsch, E. F., and Maniatis, T. (1989) Molecular Cloning : A laboratory manual. 2nd edition. 3 Vols. Cold Spring Harbor Laboratory, Cold Spring Harbor,. New York.

33. Schuler. M. A. and Zielinski. R. E.(1990) Methods in Plant Molecular Biology. Academic Press Inc. New York.

34. Seed, B. (1982) Attachment of nucleic acids to nitrocellulose and diazonium substituted supports. In genetic Engineering : Principles and Methods. (Ed : J. K. Setlow and A. Hollaender, Vol 4, : 91. Plenum Publishing, New York.

35. Southern, E. M. (1975) Detection of specific sequences among DNA fragments separated by gel electrophoresis. J. Mol. Biol : 98 : 503.

36. Ullrich, A., Shine, J., Chirgwin, J., Pictet, R., Tischer, E., Rutter. W. R. and Goodman, H. M. (1977) Rat insulin genes: Construction of plasmids containing the coding sequences. Science. 196 - 1313.

37. Wieslander, L. (1979) A simple method to recover intact high molecular weight RNA and DNA after electrophoretic separation in low gelling temperature agarose gels. Anal. Biochem. 98 : 305.

38. Williams, J. G. K., Kubelik, A. R., Livak. K. J., Rafalski, J. A and Tingey, S. V. (1990) DNA polymorphisms amplified by arbitrary primers are useful as genetic markers. Nucl. Acids Res. 18 : 6531 - 6535.

39. Zyskind. J. W. and Bernstein. S. I. (1992) Recombinant DNA: Laboratory Manual. Academic Press Inc. San Diego..

6

Plant Genetic Engineering

Biotechnology offers several potential benefits to agriculture both for the production of improved food crops and for the utilization of plants in the production of commercially important products. The technology of gene manipulation has been widely used in plant systems. Significant achievements have been made in recent years in generating transgenic plants with desired traits.

Transgenic plants have been produced for a variety of reasons such as:

1. To introduce resistance to insect pests and to various fungal and viral diseases.
2. To improve the nutritional quality (such as amino acid composition of the proteins) of food crops.
3. To increase the storage and shelf life of vegetables and fruits.
4. To increase the rate of photosynthesis.
5. To introduce new pigment colors in horticultural produce (flowers).

6. To induce herbicide resistance and for several other purposes.

The gene transfer methods in plants are different from those used with animals because of the presence of the cell wall. The initial methods of gene transfer have utilized the natural plant pathogenic bacterium *Agrobacterium tumefaciens* transformed with engineered constructs. However, this method was successful only in the dicotyledonous plants. Later publications proved that the *Agrobacterium* mediated gene transfer could be possible even in the monocotyledonous plants to which the cereals belong. Meanwhile several other methods of direct gene transfer were developed of which the most popular and successful one has been the particle gun method apart from the method of electroporation of protoplasts.Viruses have also been used for gene transfer. Basically two types of viruses, the Gemini viruses and the Caulimo viruses (cauliflower mosaic virus) are used for the transfer of genes. However, due to their pathogenic nature, restrictions on genome size and difficulties in actual gene transfer, much progress has not been made in producing viral vector based transformation systems.

Presently, the protocols for the *Agrobacterium* mediated gene transfer, gene transfer by the particle gun method and the gene transfer by electroporation of protoplasts are the most dependable and they are described here. For many gene transfer projects, an essential pre-requisite is to standardize protocols and conditions for efficient *in vitro* plant regeneration through organogenesis directly from the explant or through embryogenesis from the callus. Therefore, the methods used for plant tissue culture are described here. Important differences in hormone requirements have been observed for different plants, so adjustments will probably be necessary.

PLANT CELL AND TISSUE CULTURE

Plant cell and tissue culture is presently of great interest to molecular biologists, plant breeders and also to the horticulture industry. Tissue culture methods have become important aids to conventional methods of plant improvement. Tissue culture has been used as a method of propagation of genetically manipulated and transformed plants and also for germplasm conservation through cryopreservation etc. These methods are utilized to develop pathogen - free plants as well as in the synthesis of many secondary compounds (including pharmaceuticals).

Plant cell and tissue culture (*in vitro* culture) involves the growth and maintenance of plant tissues in a nutrient medium *in vitro* (Fig - 29 & 31). It implies the culture of protoplasts, cells, tissues, organs or whole plant culture under aseptic conditions (sterile conditions) on a nutrient medium (supplemented with plant hormones like auxins, cytokinins, gibberellins etc)of specific pH and under controlled conditions of temperature and light. The explants (any part of a plant used for initiation of culture) are generally obtained from sterile plants grown under aseptic conditions but they may also be obtained from outdoor plants. The latter require more thorough washing and disinfection before inoculation. These explants can directly give rise to multiple shoots which can be separated and grown to maturity and induced to form roots on media containing a higher auxin content (rooting media). They can also give rise to an unorganized, proliferative mass of cells called callus. Further sub-culture of the callus on media containing different combinations and compositions of phytohormones could result in the regeneration of plantlets through somatic embryogenesis or through organogenesis (shoot development). The regenerated plantlets can be gradually adapted to the green house. Synthetic seeds can be developed by entrapment of the somatic embryos in calcium alginate beads.

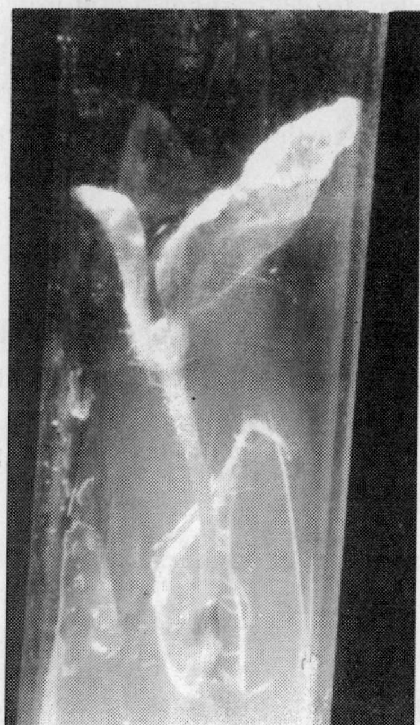

Fig. 29. *In vitro* cultured plant material.

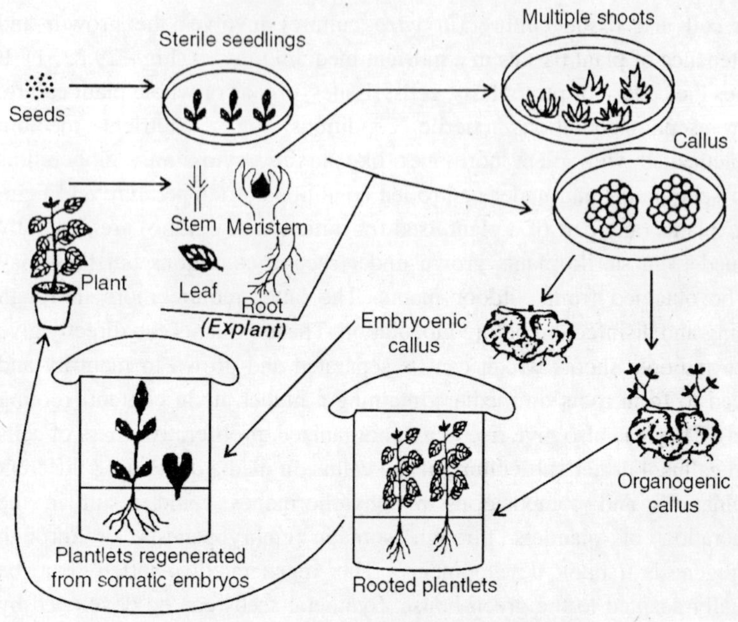

Fig. 30: Plant tissue culture.

Culture of protoplasts is also a major aspect of *in vitro* culture. Protoplasts are prepared by enzymatic digestion of cell walls. These protoplasts could be utilized for somatic hybridization and gene transfer experiments. They are then stimulated to reform a cell wall and are cultured on nutrient media for plantlet regeneration.

For efficient tissue culture, a well equipped laboratory and temperature controlled growth rooms are essential. It is also essential to adopt correct methodology and maintain aseptic environment.

Tissue culture laboratory

The minimum basic facilities required in a tissue culture laboratory are given below:

1. Clean disinfected culture rooms or incubators with controlled temperature, humidity and cool fluorescent lighting (Fig 31).
2. Clean working surfaces, continuous supply of electricity, water, gas and sufficient cabinets or shelves to store chemicals and glassware.

3. Laminar air-flow (sterile hood) work stations, optimally, these will be fitted with ultra-violet lighting for aseptic transfer and manipulations (Fig 32).

4. Spacious washing areas comprising large sinks provided with hot and cold water supply. Large plastic containers to soak the glass vessels and also for disposal of waste.

5. Instruments and tools:

   - Hot-air oven for dry heat sterilization of glassware.
   - Autoclave for steam sterilization of media and instruments (Fig 33).
   - Hot plate - cum magnetic stirrer to dissolve the chemicals for media preparation.
   - Sensitive balances for accurate weighing of chemicals.
   - Refrigerator to store chemicals, stock solutions and plant materials, and freeze (-20^{0}C, -70^{0}C.).
   - pH meter to adjust the pH of culture media.
   - Centrifuge mainly used for isolation of protoplasts.
   - Water distillation apparatus (glass) or deionizer for obtaining distilled water.
   - Filter sterilizer to sterilize heat labile liquids.
   - Dissecting microscopes for observing tissues.
   - Culture racks, trays etc.
   - Spirit lamps, sterilizers or bunsen burners to sterilize forceps, needles and scalpels used in dissection of plant material and transfer into culture vessels.
   - Hemocytometer to determine cell counts.
   - Sterile culture vessels (Fig 31) like petri dishes, magenta boxes, screw capped bottles and tubes (disposable sterile plasticware can be used).
   - Measuring cylinders, pipets, volumetric flasks, Erlenmeyer flasks and beakers.

➢ Parafilm rolls to seal the culture vessels.

Fig. 31: Tissue culture / growth room with culture vessels.

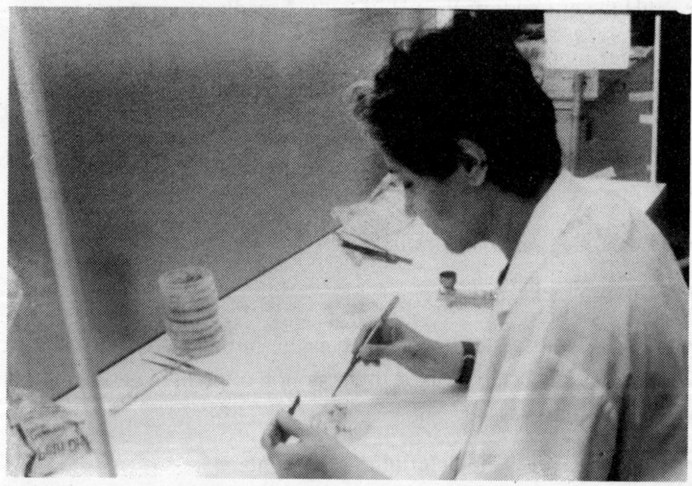

Fig: 32. Inoculation of explants on a laminar air flow bench.

Fig. 33. Autoclave. (portable)

Tissue culture media

Nutritional requirements for optimal growth of a tissue *in vitro* varies from species to species and suitable media has to be standardized in each case. Different parts of a plant (explants) can have different optimal culture media. A culture medium essentially includes inorganic salts, a carbon and energy source and vitamins. Phyto hormones and other growth stimulating chemicals are also included in the culture media to elicit efficient whole plant regeneration. The nutritional components of some plant tissue culture media are presented in the appendix (however, ready to use mixes are available today and these save a lot of time and effort in the preparation of media).

The tissue culture media are used in the liquid form to raise suspension cultures from bits of calli (with continuous shaking on a rotary shaker). However for the culturing of tissue, the media is solidified with the help of phyta-agar or phyta-gel (3 - 5 % w/v). The pH of the culture medium is adjusted to the required value (for specific media) and the medium is autoclaved. In case of the commercially available ready to use media, the powder is dissolved in water along with sugar (generally about 30 g / l w/v sucrose. Grocery-store grade sugar is adequate and less expensive!) and growth regulators if required and the pH adjusted. The phyta-agar or phyta-gel is added and auto claved.

The sterilization of media and containers etc are carried out in the auto clave is carried out at 121^{0}C (1.05 kg/Cm_2 or 15 psi) for 20-30 minutes depending on the size of the auto clave and the volume of media to be sterilized. The medium is poured into thoroughly clean sterile culture vessels (disposable petri dishes can be used) on a sterile Laminar flow bench-hood. The culture vessles with sterile media are then stored in an aseptic environment for future use. Explants or tissue are inoculated (Fig - 32)into the culture vessels under aseptic conditions on a laminar flow after a lapse of 24 - 48 h during which any bacterial or fungal contamination would show up.

Preparation of culture medium

MS basal medium (Murashige and Skoog, 1962)

For convenience, concentrated stock solutions are prepared for different categories of the constituents of MS medium. These stock solutions can be diluted according to one's need (refer to the composition of the MS medium in the appendix.)

- **Stock solution of macro-salts (10 X)**

  The macro salts should be at 10 times their final concentration (10 X).

Majors	Actual concentration mg/L	10 X (g)
$NH_4\ NO_3$	1650	16.500
KNO_3	1900	19.000
KH_2PO_4	170	1.700
Mg SO_4. $7H_2O$	370	3.700
$FeSO_4$. $7H_2O$	27.8	0.278

Make the final volume to 500 ml with d.H_2O.

Use 50 ml of stock to prepare 1000 ml of culture medium

- **Stock solution of micro salts (100 X)**

Minors	Actual concentration mg/L	100 X (mg)
H_3BO_3	6.2	620
$MnSO_4 . 4H_2O$	22.3	2.230
$ZnSO_4 . 7H_2O$	10.3	1.03
$Na_2Mo . O_4 . 2H_2O$	0.25	25
$CuSO_4 . 6H_2O$	0.025	2.5
$CO\ Cl_2 . 6H_2O$	0.025	2.5 mg

Dissolve in 100 ml dH_2O

Use 1ml for preparing 1000 ml of culture medium.

- **Stock solution of calcium chloride (10 X)**

Add 4.400 g to 100 ml of dH_2O

Use 10 ml for preparing 1000 ml of culture medium (an equivalent of 440 mg/L).

- **Stock solution of potassium iodide (100 X)**

Add 83 mg to 100 ml dH_2O

Use 1 ml for preparing 1000 ml of culture medium (an equivalent of 0.83 mg/L).

- **Stock solution of myoinositol (10 X)**

Dissolve 1000 mg in 50 ml dH_2O

Use 5 ml to prepare 1000 ml of culture medium (an equivalent of 100 mg/L).

- **Stock solution of Na_2 EDTA (10 X)**

Dissolve 373 mg in 50 ml of dH_2O

Use 5 ml to prepare 1000 ml of culture medium (an equivalent of 37.3 mg/L).

- **Stock solution of vitamins (100 X)**

Vitamins	Actual concentration (mg)	100 X (mg)
Glycine	2	200
Thymine HCL	0.1	10
Pyridoxine HCL	0.5	50
Nicotinic acid	0.5	50

Dissolve in 100 ml dH_2O

Use 1 ml to prepare 1000 ml of culture medium

- **Stock solutions of hormones**

The stock solutions of hormones are optional and can be used with any medium apart from MS medium. All the stocks are prepared in the proportion of 10 mg/10ml of absolute ethanol and stored in the freezer at -20^{0}C.

Carbohydrate source

The carbohydrate source of MS medium is sucrose. 30 g / L sucrose is the actual amount used generally. To obtain a solid medium, 4 - 5 g / L agar is added to the liquid medium.

Procedure: To prepare 1 liter culture medium

1. Dissolve 30 g sucrose in 200 ml of double distilled H_2O in a flask and set aside

2. Take 100 ml dH_2O in another flask and add the required amounts of stock solutions in a sequence to it. (If the ready to use powder is being used, simply mix the powder along with the sucrose in distilled water and make it upto 1 liter).

3. Mix both the solutions and make up to 1 liter with dH_2O in a large autoclavable flask. Adjust the pH to 5.8 with drops of 0.1 N HCl or 0.1N NaOH by using the pH meter.

   - *Optional: Add 0.5 mg/L MES [2-(N-morpholino) ethane sulfonic acid] to provide buffer to prevent pH changes during plant cell growth.*

4. If a solid medium is to be prepared then add the required quantity of phyta-agar or phyta-gel.
5. Seal the mouth of the vessel tightly with aluminium foil, and fix an autoclave tape around it. Autoclave the medium at 121^0C (1.05 kg/cm^2) 15 psi for the required time (ranges from 20 - 40 min, with larger volumes of media requiring more time.
6. Remove the vessel and let it cool. Open the sealed packaging of the sterile disposable petri dishes and keep them in the sterile hood. Other sterile or autoclaved vessels can also be used including food jars. Swirl the medium well and pour it into the culture vessels to the desired appropriate levels. Cover the vessels immediately . Do not disturb them till they cool down and the medium solidifies. Store these in their sterile wrappings in a clean place.

Sterilization and inoculation of plant material

Plants and seeds carry a wide range of microbial contaminants like fungi and bacteria. Hence the plant parts brought from out door plants or seeds (from which sterile plants can be grown in sterile containers) need to be thoroughly sterilized in disinfectant solutions (refer appendix) before the transfer onto the respective culture medium contained in culture vessels. Precautions are taken by the operator to wear a sterile mask over his nose and mouth, a sterile head-gear and lab coat before starting the manipulation. The hands are rinsed with 95% ethyl alcohol and the laminar flow bench area sterilized by switching on the U.V light for 30 min before swabbing and spraying the entire area with 95% ethanol. The scalpel and forceps are kept in a beaker dipped in ethanol. For inoculation of sterile plant material into the culture vessel (Fig - 32), the forceps and scalpel are flamed and then used to pick up the plant material and transferred into the vessel, before closing the lid immediately.

Incubation and maintenance of the cultures

After the inoculation of explants, the culture vessels are tightly sealed with parafilm to prevent evaporation and to preserve the sterile conditions. They are then incubated in a sterile, temperature and humidity controlled room with provision to regulate light and dark periods according to the need of the specific plant species and experimental conditions (Fig-31). The explants either develop shoots directly (single/multiple) or produce a callus (Fig-30).

The shoots may be transferred to a rooting medium (generally with higher auxin levels) for the development of roots and later transferred to pots kept in a mist chamber or green house for acclimatization of the regenerated plantlets. The calli on the other hand require sub-culture to media supplemented with several combinations and concentrations of phytohormones. The calli then start to either develop shoots through organogenesis when they can be transferred to a rooting medium or develop somatic embryos through embryogenesis which later grow into whole plantlets. Protoplasts can also be cultured in liquid media and later on transfered to solid media when they start to develop into calli and later plantlets.

In vitro culture of plant tissues

Plant material

The explants can be obtained from out door plants or from sterile plants grown *in vitro* from sterilized seeds.

Method to raise sterile plants

Wash the dry seeds with tap water, soak in 5% detergent solution (v/v) for 15 min and rinse well with distilled water. Carry them to the sterile hood and rinse with 70% ethyl alcohol. Transfer the seeds to a sterile flask and pour 0.1% $HgCl_2$ solution (w/v) to immerse them. Decant 15 min later after thorough shaking and rinse 4 times with sterile dH_2O. ($HgCl_2$ is toxic and must be handled with gloves ! An alternate method of sterilization is to use 25 - 50 % commercial bleach.). Arrange the seeds on the surface of the culture medium (basal medium) in culture vessels. Incubate the culture vessels at 25 - 28^{0}C or room temperature till the seeds germinate and grow into long seedlings. The seeds from different plants require light or dark to germinate. The explants comprising the stem tip, root, cotyledons, leaf, epicotyl, hypocotyl or the shoot meristem can then be obtained from these aseptically grown seedlings for the culture.

Procedure

Take out each instrument from the alcohol and flame it before use. Take precautions to prevent the alcohol in the beaker or your fingers from catching fire!

1. Clean the sterile hood and sterilize it with ethanol as described above. Keep the culture vessels ready.

2. Wear a clean lab coat and mask. Clean the hands with alcohol and dip all scalpels, forceps and needles in a beaker with ethanol. Make sure that the burner is clean and ready with a flame.

3. If the explant material is from outdoor plants, sterilize it according to the method explained for the seeds but use only the commercial bleach and place it in a sterile petri dish. Cut the explants out from the sterile plants and get ready to inoculate them into the culture vessels.

4. Quickly transfer the explant onto the medium with a forceps and close the mouth of the vessel. Seal the vessel with parafilm. To counteract contamination problems encountered with explants from non-aseptic plants, antibiotics can be included in the medium: 80 µg / ml ampicillin, 10 µg / ml rifampicin, 50 µg / ml benomyl etc.

5. Incubate the culture vessels in a temperature controlled sterile room at 25 - 27^{0}C with cool fluorescent illumination.

6. Check regularly for callus development or direct shoot formation and remove contaminated vessels immediately.

7. Sub -culture the callus onto fresh media after 2 - 3 week intervals to multiply it or to induce plantlet regeneration. Plantlet regeneration can occur through organogenesis or embryogenesis.

8. Shoot meristems or axillary buds or nodal explants with axillary buds can be inoculated onto culture media to induce multiple shoots which can then be rooted on rooting media to obtain efficient plantlets without going through the callusing stage.

9. Carrot root discs can be used as explants for callus induction routinely by inoculating them onto MS medium supplemented with 1mg/L 2,4-D.

Protoplast isolation and culture

Isolation of protoplasts from leaf mesophyll cells

Protoplasts can also be isolated from suspension cultures)

Wear a clean lab coat and mask. Clean the hands with alcohol and dip all scalpels, forceps and needles in a beaker with ethanol. Make sure that the burner is clean and ready with a flame.

Requirements:

- Leaves from 15 - 20 d old pea plants growing in a glass house.
- Protoplast culture medium (see appendix)

  C P W Solution (cell - protoplast washing)

  KH_2PO_4 27.2 mg

  KNO_3 101 mg

  $CaCl_2.2H_2O$ 1480 mg

  $MgSO_47H_2O$ 246 mg

  KI 0.16 mg

  CuSO4. $5H_2O$ 0.025 mg

  Bring up to 1000 ml with dH_2O . Adjust the pH to 5.8 with 0.2N KOH or 0.2N HCl. Prepare CPW 13M by dissolving 13 g mannitol in 100 ml CPW solution.

  Prepare CPW 21 S by dissolving 21 g sucrose in 100 ml CPW solution.

- **Enzyme solution**

  Prepare the enzyme solution with CPW 13M by adding 2% cellulase (Onozuka R10) and 0.5% macerozyme or pectinase and adjusting to pH 5.5. Pass this solution through a membrane filter (0.45 μm) to sterilize it and use immediately. This solution can be stored in a refrigerator but should be filter sterilized before use.

Procedure

Refer Fig-34, to understand the procedure of isolation and culture of protoplasts.

1. Surface sterilize the leaves by immersion in 70% ethanol for one minute (in the sterile hood) and rinse in 10% sodium hypochlorite and tween-80 for 30 seconds. Rinse thoroughly with sterile distilled water.

2. Peel off the lower epidermis with a fine forceps or cut the leaves into thin strips and immerse them in a petri dish containing CPW 13M solution for 30 minutes for plasmolysis.

3. Remove the CPW 13M solution with a pasteur pipet and replace it with the enzyme solution.

4. Seal the petri dish with parafilm and incubate it at 25^0C in the dark for 4 -16 hours on a shaker (at slow speed of 50 rpm) or shake gently at hourly intervals.

5. Observe the petri dish under a microscope for the release of protoplasts.

6. Remove the protoplast solution with a Pasteur pipet and filter it through a 60-80 μ m stainless steel sieve to remove the debri (on a laminar flow bench).

7. Transfer the protoplast solution to a screw capped centrifuge tube and centrifuge at 100g for 4 min. Decant the supernatant.

8. Suspend the pelleted protoplasts in CPW 21S solution and centrifuge for 6 min at 200 rpm. The viable protoplasts will float at the top of the solution in a dark green band. They can be observed under the microscope. The viability of living protoplasts can be tested by Evan's blue stain (Dissolve 250 mg Evan's blue stain in 100 ml of 12% mannitol). Viable protoplasts do not stain.

9. Collect the viable protoplasts and suspend in CPW 13M and centrifuge twice with CPW 13 M washings.

10. Use a hemocytometer to count the number of cells in 10^{-4} ml and adjust the protoplast density to 0.5 X 10^5 to 1×10^5 ml/ litre with liquid culture medium.

11. The protoplasts can be incubated in sealed petri dishes in the form of small drops or a thin layer of the protoplast solution under low-light or dark conditions at 25 - 28^0C.

12. Bergmann's cell plating technique may be followed where 2 ml aliquots of isolated protoplasts (10^3 - 10^5 cells ml^{-1}) are mixed with an equal volume of agar nutrient medium (of temperature not exceeding 45^0C) and poured into petri dishes which are then sealed and incubated in an inverted position (after the solidification of agar) at 25 - 28^0C under light / dark conditions. These can be studied under the microscope for further development of protoplasts.

13. Alternately, the liquid culture medium with protoplasts can simply be poured over the agar medium in the petri dishes. The medium can be modified to suit different materials.
14. Protoplasts develop cell walls after 2 - 3 days in culture medium, start to divide in 2 - 7 days and develop into multicellular colonies after 2 - 3 weeks in culture. Large colonies will be visible after 2 - 3 weeks.

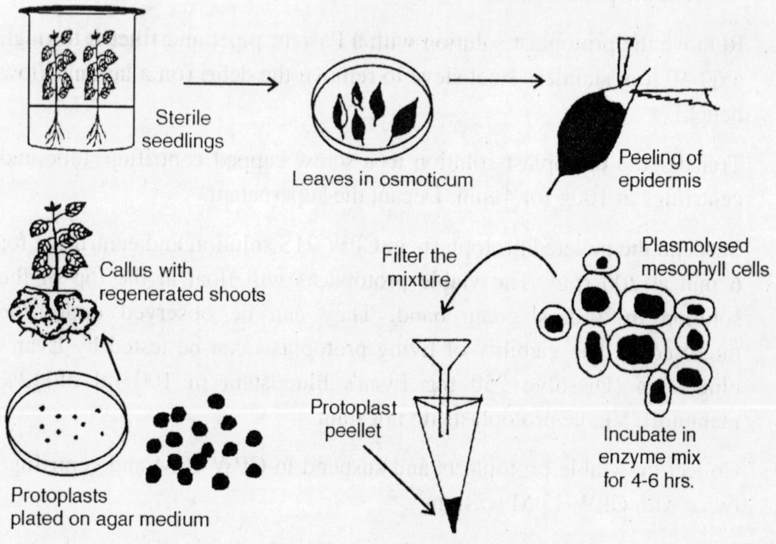

Fig. 34: Isolation and culture of protoplasts.

Protoplast fusion (somatic cell hybridization)

Poly ethylene glycol method :

Poly ethylene glycol (PEG) method of protoplast fusion is one of the easiest ones. PEG acts as a molecular bridge connecting the protoplasts and Ca^{2+}ions bind the membrane surfaces together.

Requirements

- Protoplasts from the two parents suspended in enzyme solutions
- Protoplast culture medium
- **Washing solution**

500 mM glucose

0.7 mM KH_2PO_4. H_2O

3.5 mM $CaCl_2.H_2O$........pH 5.5

- **Poly ethylene glycol solution (PEG)**

  PEG (1500 - 2000 M.W) 50 g

  Glucose 1.8 g

  Ca Cl_2. $2H_2O$ 150 mg

  K_2 HPO_4 12.0 mg

  Dissolve glucose, $CaCl_2$ and K_2 HPO_4 in 80 ml of dH_2O. Add PEG and bring upto 100 ml. Autoclave for 20 min to dissolve it.

- **Eluting solution**

  300 mM glucose

  50 mM glycine

  50 mM $CaCl_2.H_2O$

  Adjust pH to 9 - 10.5.

- **0.1% calcofluor**

  Prepare 0.1% calcofluor white in 0.4 M sorbitol.

- Silicon 200 fluid

Procedure

1. Mix the two protoplasts in equal proportions.
2. Transfer to a screw cap centrifuge tube and centrifuge at 50 x g for 5 minutes.
3. Remove supernatant with pasteur pipette and discard. Wash the protoplast pellet with 10 ml washing solution and centrifuge. Finally suspend the protoplasts in 0.6 ml of PEG solution (added in drops). Cap the tube and incubate at room temperature for 40 min by shaking very gently. Alternately for observing the agglutination, the PEG solution and protoplast can be mixed in a depression slide and observed after placing a cover slip.

4. Pour 1 ml of eluting solution into the tube, centrifuge at minimum speed and discard the supernatant and continue the second, thrid, fourth and fifth washings with protoplast culture medium.

5. Culture the protoplasts in the petri dishes as explained in the earlier experiment. The protoplasts can be screened under a microscope for the hybrid fusion cells among the unfused parental types, homokaryons etc. Biochemical markers can be used to identify true somatic hybrids. Flow cytometric analysis is also used increasingly. The nature of somatic hybrids and cybrids is assessed on the basis of morphological and karyological features of the fusion products.

Preparation of synthetic seeds

Somatic embryogenesis has a potential application in plant improvement. Encapsulation of somatic embryos enables them to be sown under field conditions as synthetic or artificial seeds (Fig - 35). Synthetic seeds consist of somatic embryos enclosed in a protective coating and are proposed as a low - cost - high volume propagation system. After encapsulation, the embryo conversion frequency and development of normal plant are routinely tested.

Requirements

- Embryogenic callus with well differentiated somatic embryos from any explant culture**. Sodium alginate solution**

  Prepare 4% w/v sodium alginate (can be prepared in MS medium). Autoclave.

- **Calcium chloride solution**

  Prepare 100 m M $CaCl_2 . 2H_2O$ with dH_2O. Autoclave.

- Liquid nitrogen
- Culture vessles with MS basal medium (solid)

Procedure

1. Isolate the somatic embryos and mix with the sodium alginate solution.
2. Take a wide bore (2 mm) pipette or glass tube with wide bore and suck the embryos along with sodium alginate.

3. Drop the embryos one by one distant from each other in a beaker containing the Ca Cl_2 solution. The Ca Cl_2 reacts with sodium alginate to form a bead composed of calcium alignate enclosing the somatic embryo.
4. Allow the embryos to remain in the solution for 20 minutes by gentle shaking once in every five minutes so that proper beads are formed.
5. Remove the beads seal in cryogenic tubes and place them in liquid N_2 for storage.
6. Test the beads for plant conversion by inoculating a single bead onto a culture tube slant following its germination response.

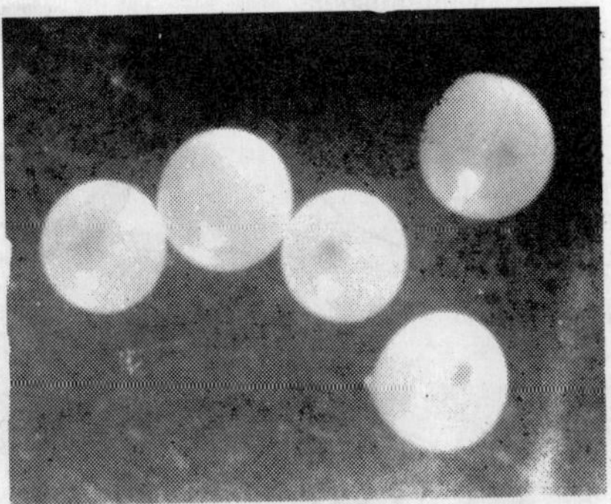

Fig. 35: Synthetic seeds.

GENE TRANSFER AND EXPRESSION IN PLANTS

Genetic transformation of plants using *Agrobacterium* system

Plant genetic transformation involves the transfer of foreign genes into plants. To facilitate the transfer of genes, a series of vectors have been developed based on the *Agrobacterium* Ti-plasmid transfer system.

The soil bacteria *Agrobacterium tumefaciens* and *A. rhizogenes* have the natural ability to transfer genes into plant cells. Wild type *A tumefaciens* induces tumors at sites of wounds of dicotyledonous plants. This process depends on the presence in the bacterium of a tumor inducing (Ti) plasmid. The presence of phenolic compounds exuded from the wounded cells induces transcription of genes in the virulence (vir) region of the plasmid that are involved in the excision of a part of the plasmid known as the T (transferred) region. The T-DNA is then transferred to the plant cell and stably integrated into its genome by an unresolved mechanism (Fig. 36). The wild type Ti - plasmid contains genes that code for the production of unusual amino acids called opines and also for plant hormones (auxins and cytokinins). The production of excess hormones causes a gall to form at the site of infection and the bacteria utilize the opines (as a carbon and nitrogen source). Similarly, *A. rhizogenes* transfers Ri T-DNA that induces the formation of transformed hairy roots at the infection sites.

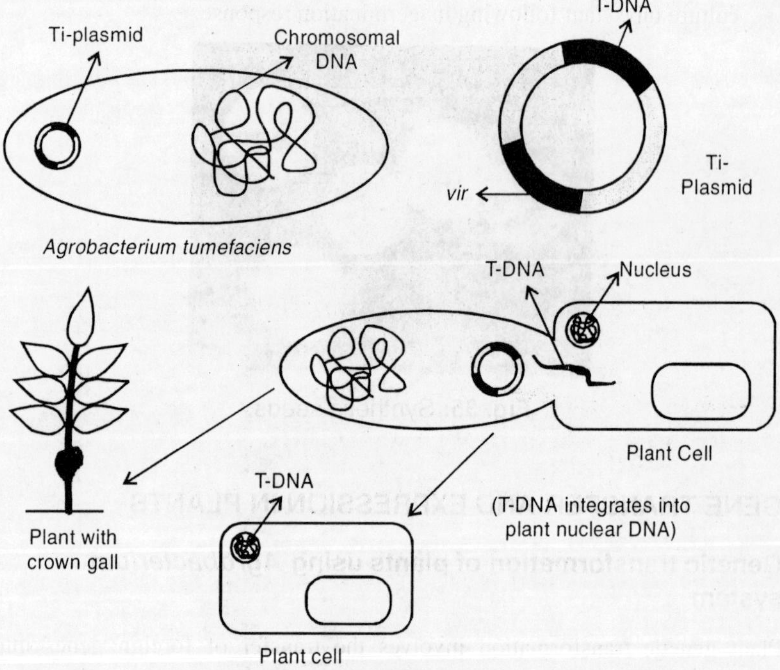

Fig. 36: *Agrobacterium* infection.

In the vectors developed from the Ti - plasmid system, many of the wild type genes were excised ('disarmed' vectors) and replaced by genes of interest to be introduced into plants. Removal of genes that code for phyto - hormone overproduction and opine production results in normal transformed plants without any tumours. Foreign genes of interest were introduced in place of the excised genes. Any inserted foreign gene must have an appropriate promoter for successful expression in the host plant. Also, good protocols of *in vitro* plant regeneration have to be developed so that the transformed plants can be grown to maturity. However, *Agrobacterium* does not transform all dicotyledonous plants with equal efficiency and does not normally infect monocotyledonous plants. However, there have been several reports on successful transformation of monocots recently.

The specific characteristics of the disarmed vectors used in genetic transformation are:-

1. Vectors could be co-integrate (virulence region and the T-DNA portion in the same plasmid) or binary (in different plasmids).

2. Modified T-DNA: T-DNA from which the oncogenic (tumor inducing) genes have been removed and the foreign gene with a promoter inserted in their place flanked by the border sequences.

3. To recover the transformants after the transformation, specialized tissue selection media have to be designed based on the marker gene. The constructed vectors usually have selectable marker genes (like NPT I I which codes for kanamycin, the herbicide resistance gene, *bar* etc). Generally, most constructs have the GUS gene, a scoreable marker for which an assay can be carried out for the transient or stable expression.

The Agrobacterium mediated methods of transformation are explained below (Horsch et al. 1989)

Induction of crown gall or hairy roots using wild strains of Agrobacterium

Plant material

- Carrot roots.

Bacterial strains

- Wild strains of Agrobacterium tumefaciens and of A. rhizogenes. *(These can be procured from any culture type collection centers).*

- Medium for Agrobacterium culture and maintenance (liquid medium):

Mannitol 2 g

Yeast extract	10 ml
$K_2\ HPO_4$ (2%)	10 ml
$K\ H_2PO_2$ (2%)	0.8 ml
$Mg\ SO_4$ (1M)	0.4 ml

Make up the volume to 1 litre with distilled water and autoclave. For solid medium, add 5 g of agar.

Solutions :

- MS basal medium with 300 μg / ml of the antibiotic, carbenicillin / cefotaxime.

  The antibiotic is added to prevent contamination by other bacteria.

- Stock solution of antibiotic

  Dissolve 1.0 g of carbenicillin or cefotaxime in 20 ml of water and filter it through 0.2 μM membrane filter to sterilize it. Store it at 4^0C.

Procedure

1. Prepare fresh agar slants of the bacterial medium and inoculate loopful of cells from the stock culture onto the slants (a loopful on each slant).
2. Allow the cells to grow for 48 hrs at 25^0C.
3. Autoclave 20 ml of liquid bacterial medium, let it cool to 45^0C and add the antibiotic. Pour the medium into sterile petridishes and allow them to solidify on the sterile hood.
4. Inoculate a loopful of cells from the agar slants into the liquid medium and grow the cultures overnight or for 48 hrs on a shaker at 150 rpm at 25^0C.
5. Dip the carrot in ethanol and flame it. Remove the skin aseptically and slice it into discs.
6. Place the carrot discs in the petridishes containing the medium.
7. Place 20 μl of the liquid bacterial culture on each disc under aseptic conditions.
8. Incubate the petridishes under light/dark in the culture room at 25^0C.

9. Observe the plates for the development of galls / roots alround the discs from the second or third week. Record the response under light and dark incubation.

10. This technique can be modified when using explants from aseptically grown sterile plants. In such a case, place the explants in a sterile petridish containing 15 ml of MS liquid medium(without carbenicillin) and 500 µl of overnight grown liquid bacterial culture. Incubate for 48 to 72 hrs under light at 25^0C. Seal with parafilm Wash the explants thoroughly with sterile dH_2O and place them on MS agar medium supplemented with carbenicillin. The galls/roots appear around second/third week.

Leaf disc transformation t echnique using Agrobacterium constructs: (Horsch et al. 1989)

Plant material

Any dicotyledonous plant. But preferably a member from Solanaceae because good callusing and plantlet regeneration protocols have to be established for the experimental plant material before the transformation experiments are taken up.

Agrobacterium strains

Agrobacterium strains with constructed vectors are used. The vectors are generally cointegrative or binary type. Both may have scoreable and / or selectable marker genes apart from a promoter(e.g the cauliflower mosaic virus promoter) for expressing the marker genes and the genes of interest.

Selection medium

Prepare selection media plates with 300 µg/ml cefotaxime and 50 µg / ml kanamycin (if the selectable marker gene NPT II is a part of the construct) or 1 to 3 mg / ml phosphinothricin (if the *bar* gene is present in the construct).

Procedure

1. Germinate the sterilized seeds of the experimental plant on MS basal medium and incubate at 25^0C till 4 - 5 leaves develop on the seedlings.

2. Grow the *Agrobacterium* strain in 10 ml of LB or any specified liquid medium at 25^0C on a rotary shaker at 200 rpm for 48 hours (in a 100 ml flask).

3. Remove the leaves of the sterile seedlings aseptically and place them in a sterile petri dish.

4. In another sterile petri dish, pour 20 ml of MS basal (liquid) medium and add 200 µl of the 48 hours bacterial culture to set up the co-cultivation.

5. Cut the leaves into 1 cm discs and float them upside down on the MS + bacterial culture medium. Co-cultivate them for 72 hrs at 25^0C under light. Rinse with sterile distilled water and transfer them to selection medium.

6. Incubate the plates at 25^0C and observe for callus formation.

7. Sub-culture the callus onto the selection medium (addition of cefotaxime is not necessary from this stage onwards).

8. After considerable growth of the callus is obtained, transfer it to appropriate regeneration medium to enable them to regenerate into plantlets.

9. Confirm the transformation by GUS assay for transient expression of the GUS gene (if the gene is a part of the construct) and later subject the regenerated plantlets to Southern blotting and hybridization to confirm the integration of the foreign DNA.

Genetic transformation of plants using direct methods

When earlier attempts to transfer genes to monocotyledonous plants did not succeed (they have however succeeded now!), alternative methods were developed to transfer foreign genes directly into plant cells using physical rather than biological means. Two such direct methods are

1. Electroporation

2. Particle gun method through microprojectile bombardment.

Electroporation involves the use of an apparatus where in electric impulses are transimitted to protoplasts (Fig - 37). A high concentration of plasmid DNA containing a cloned gene is added to a suspension of protoplasts and the mixture subjected to an impulse with an electrical field of 200 to 600 v / cm. The protoplasts are then nurtured and cultured on selection media to isolate the transformed plants.

A much simpler procedure, where one need not isolate protoplasts and deal with very sensitive culture conditions is the micro-projectile mediated gene transfer. The foreign DNA is coated onto tiny metal beads (1 μM in diameter) made of tungsten or gold and these particles (micro projectiles) are precipitated on the macro projectile which is then fired from the 'gun' with velocities of about 430 meters/sec. The targets into which the DNA is shot vary from suspension cultures of embryonic cells plated on filters, calli, intact leaves (even maize kernels), pollen grains or apical meristems.

Generally the transformed cells are cultured on selection media to eliminate all non-transformed cells. The transient expression of the GUS gene is tested in the transformants if the gene is present in the construct. The presence of the foreign gene or any marker gene is confirmed by Southern blotting and hybridization with a complementary labeled probe.

The two direct methods of gene transfer viz: Electroporation and micro projectile bombardment are described here followed by the methods of analysis to be carried out on the putative transformants for confirmation of gene transfer.

Gene transfer through electroporation (Fromm et al. 1986)

Protoplasts from mesophyll cells or suspension cultures can be used for electroporation. The vector DNA carrying the foreign gene, markers and promoter is isolated from the host bacterial cells and used for the transfer.

Solutions

- Electroporation buffer:

  5 mM 2-(N-morpholino) ethane sulfonic acid (MES) buffer (pH 5.8)

  0.5 M mannitol

  70 mM KCl

  1% poly ethylene glycol.

Procedure

1. Isolate the protoplasts (follow the procedure given earlier).
2. Adjust the density of the protoplasts to 5 x 10^7 /ml.
3. Wash once in the electroporation buffer by centrifugation and collect the pellet.
4. Suspend the pellet in a little buffer to the same concentration as above. Heat -shock the protoplasts in the tube for 5 min at 45^0C in a water bath and then cool on ice.
5. Set up the electroporation system to a capacitance of 800 µF, resistance of R3 (48 Ohm) and charge of 106 volts.
6. Add 20 µl (1µg/ml) vector DNA and 125 µl of PEG solution to 0.25 ml of protoplast suspension in a 15 ml polystyrene tube, mix gently and pipet into a sterile cuvette placed on ice. Insert the electroporation chamber into the slider.
7. Carry out the electroporation and note down the actual pulse length in milliseconds and the voltage.
8. Electroporate protoplasts without vector DNA and treat it as a negative control.
9. Cool the electroporated protoplasts on ice for 10-30 min and culture on the standard protoplast culture medium solidified with 1.2% agarose. Examine the protoplasts for the development of cell walls in 2-3 days and for cell division and micro colony formation in 7-10 days. Transfer the tissue three weeks after the electroporation to selection media (containing the specific antibiotic with respect to the selectable marker gene) and monitor the growth.
10. The expression of the GUS gene (when present on the vector DNA) may be determined by the GUS assay of transformants. The further confirmation of foreign gene transfer can be made by the Southern blot and hybridization with a specific probe (complementary to the foreign gene).

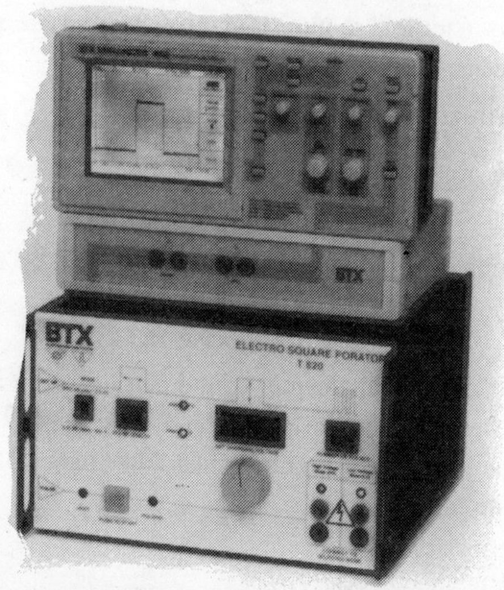

Fig 37: Electroporation apparatus

Gene transfer through micro-projectile bombardment (Klein *et al.* 1987)

Microprojectile bombardment carried out by a biolistic or particle gun, is a very powerful method of direct gene transfer (Fig - 38). It is a relatively simple and easy technique and several cells may be transformed with one shot. However, the equipment is very expensive. Different systems like meristems, pollen, calli or suspension cultures can be used for the microprojectile delivery of foreign DNA.The method described here is specific to multiple shoot clumps but can be suitably modified for other materials. The vector DNA used for transfer contains the actual gene, a promoter and selectable and/or scoreable marker genes. The plasmid DNA is isolated from the host bacterial cells and used for the transfer.

Requirements

Carry out all the procedures on the laminar flow sterile bench and maintain aseptic conditions]

- Biolistic Particle delivery system (Fig-38) (the most widely used one is PDS-1000 of Dupont's)
- Tungsten or gold particles (0.9 - 1.2 μm)

- Macrocarriers (sterile), rupture disks (of appropriate pressure), sterile stopping screens.
- Plasmid construct DNA (1 μ g / μ l).
- 2.5 M $CaCl_2$.
- 100% ethanol
- 0.2 M spermidine

Preparation of the sample

Arrange the plant tissue in the center of the petridishes in a circular area of 2.5 cm. A thin layer of suspension culture cells can be collected onto a millipore filter pad or onto a premoistened filter paper placed on an agar medium to provide support.

Procedure

1. Sterilize the tungsten or gold particles by soaking in 1 ml absolute ethanol for five min. Sonicate on ice for 5 min with a continuous pulse using a 20 % duty cycle at level 2 output.
2. Pipet out 50 μl of the mixture immediately into a microcentrifuge tube. Centrifuge at 12,000 rpm for 5 min.Discard the supernatant and wash the particles thrice with autoclaved water and resuspend in 320 μl sterile distilled water.Use 50 μl of microprojectiles or multiples according to the number of shots planned (bombardments).A single shot requires 20 μl of microprojectiles.
   - For 50 μl of microprojectiles, it would be necessary to use 15 μl of DNA
   - (1 μg / μl) and 25 μl of 0.2 M spermidine and the same amount i.e 40 μl of
   - 2.5 M $CaCl_2$..
3. Add the DNA, spermidine and the calcium chloride to the microprojectiles and vortex for 5 min. Incubate on ice for 15 min.
4. Centrifuge at 12,000 rpm for 5 min and discard the supernatant. Wash the DNA coated microprojectiles with 1 ml 100 % ethanol and resuspend them in the appropriate amount of ethanol.

- *1 shot requires 10 μl of the microprojectile-ethanol mixture. Therefore take appropriate amount of the DNA and microprojectiles and work with them and finally suspend the DNA coated particles in an appropriate amount of ethanol*
- *Load these on each macrocarrier.*

5. Dip the macrocarriers individually in isopropanol and place them on a sterile filter paper to dry. Load 10 μl of these DNA laden tungsten particles on each macro-carrier.

6. Load the rupture disc into the acceleration tube, fix the stopping screen and macrocarrier assembly and place the petridish with the sample at an optimum distance and close the door shut.

7. Press the vacuum switch till it reaches 23- 25 inches Hg and immediately press the vacuum hold switch. Press the fire switch and watch the Helium (He) gas fill up and the pressure increase to the required limit when the rupture disk bursts. Release the fire switch immediately after the rupture disk bursts and the gun fires onto the macrocarrier which is stopped by the stopping screen while the DNA coated microprojectiles are shot into the target tissue.

8. Bombard each sample twice and transfer the material to fresh sterile petridishes. Scrape up the cells immediately and immerse in fresh sterile culture medium if using suspension culture material. If the construct has the GUS gene, a GUS assay may be carried out 48 hrs after the bombardment.

9. After two to four weeks, transfer the bombarded material to selection media if the transferred construct has the selectable marker gene (the selection component depends on the selectable marker gene e.g. addition of herbicide if the bar gene is present).

CONFIRMATION OF THE TRANSFORMANTS

The transient expression of the engineered gene is analyzed by the GUS assay (if the GUS gene is present in the construct). However, the analysis for the presence and stable integration of the target genes can be carried out by the methods of PCR, Dot blot and Southern blots by using a complementary probe (these methods have been explained in chapter-5). The presence of the marker genes like the GUS gene and the *bar* gene can also be confirmed by these methods.The stable integration of a gene is generally followed by its expression (of a protein). The analysis for gene expression can be carried out by the protein hybridization method known as the Western blot (refer chapter-5).

Fig 38: PDS 1000 particle gun

GUS assay :Method for transient β - glucuronidase (GUS) gene expression by enzyme assays

(Histochemical and fluorometric assays]

The substrate used to study the GUS gene expression is 5-bromo-4-chloro-3-indolyl glucuronide (X-gluc) which gives a blue precipitate at the site of enzyme activity. Whole plants, callus, suspension culture cells or protoplasts can be used .

Histochemical assay (Jefferson, 1987)

Requirements

- **GUS histochemical staining buffer**

  100 mM sodium phosphate, pH 7.0

  100 mM sodium EDTA

  5 mM potassium ferricyanide

  5 mM potassium ferrocyanide

  0.5 % Triton X- 100

  Filter sterilize the buffer and store in sterile 50 ml tubes at -20^{0}C.

Staining solution

Add the substrate X- glucuronide (5-bromo-4-chloro-3-indolyl -B-D-glucuronic acid) to the histochemical buffer just before use in the proportion of 1 mg X- gluc to 1 ml of the buffer. A stock solution of X-gluc can be prepared by dissolving 100 mg in 4 ml of 50 % ethanol and stored at -20^{0}C. Appropriate amount can be added to the buffer just before use.

Fixative :(optional)

0 1 % (v / v) glutaraldehyde in 25 mM sodium phosphate buffer.

Procedure

1. Fix the putatively transformed plant tissue in the fixative at pH = 5.6 for 30 - 45 min at room temperature or use it straight away for staining (whole plant parts, callus, stem sections or protoplasts). A control should also be set up (i.e a non-transformed plant tissue) for comparison.

2. Wash the tissue in the phosphate buffer solution about 5 - 6 times.

3. Incubate the tissue in the staining solution by completely covering it and pull a vacuum for 10 min. Incubate in the dark at 37^{0}C overnight or for 24 hrs.

4. Rinse the tissue twice for 5 min each in phosphate buffer until the tissue shows an intense blue color. Pass the material through an ethanol series

for chlorophyll removal and clear visibility of the blue regions.The material can be stored at room temperature in 100 % ethanol.

5. Score the blue sectors. GUS staining (activity) can be seen in young leaves, sections of fully expanded mature leaves, callus, multiple shoot clumps, vascular bandles and parenchyma cells. Stems show good staining but roots, petals and flowers show a weak blue colour. The staining confirms the integration and expression of the marker gene adjacent to which the actual gene is located and therefore its integration is confirmed (Fig - 39).

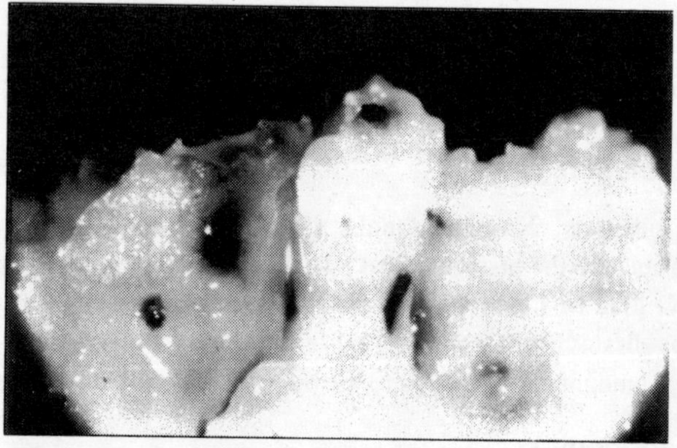

Fig 39: Transient GUS expression (blue staining) in callus.

Fluorometric assay (semi-quantitative assay) : (Jefferson, 1987)

This test is based on fluorescence. The enzyme produced by the GUS gene hydrolyzes 4-methyl-umbelliferyl-β -D-glucuronide (MUG) and produces 4-methyl-umbelliferone (4 MeU), generating a blue fluorescence from which quantitative measurements can be made.

Requirements

- **Extraction buffer**

  50 mM sodium phosphate (pH 7.0)

  10 mM EDTA

  0.1 % (v/v) Triton X- 100

  10 mM 2-mercaptoethanol

  0.1 % (w/v) sarcosyl

- **GUS assay buffer**

  50 ml GUS extraction buffer

  22 mg 4-methyl-umbelliferyl-B-D-glucuronide (MUG).

  Store at 4^0C. Prepare fresh every two weeks.

- 0.2 M Na_2CO_3

Procedure

1. Grind 50 mg of the sample (the transformed and a non-transformed negative control plant material) in 100 µl of the extraction buffer in an eppendorf tube by using a mini pestle.

2. Centrifuge for 5 min at 5000 rpm at room temperature and transfer the supernatant to another tube to proceed with the GUS assay.

3. Calculate total proteins from the sample before proceeding with the GUS assay..

4. Add 100 µl of the GUS assay buffer to the supernatant and incubate at 37^0C for 1 hr. Terminate the reaction by adding 150 µl 0.2 M Na_2CO_3.

5. Read the fluorescence and subtract the blank value from it. The specific activity (units / ng protein / min) of GUS in the tissue can be calculated based on the total proteins and incubation time.

SELECTION OF STABLY TRANSFORMED PLANTS

Transfer the *in vitro* regenerated transformed plants (R_o plants) to the greenhouse and grow them to maturity. Allow them to self pollinate and harvest the seeds. Raise the R_1 plants. Study the ratios by scoring the phenotypes of the germinated seedlings and calculate the number of independently segregating genes (with reference to a trait). One gene can produce a 3 : 1 ratio, two genes can produce a 15 : 1 ratio and three genes can produce a 63 : 1 ratio. The R_o plants can be cross pollinated with non-transformed control plants and the next generation analyzed to produce a 1: 1 ratio in case of one gene, a 3: 1 ratio for two genes and a 15 : 1 ratio for three genes.

In order to rule out any escapes or artifacts, the transgenic plants should be grown for several generations. It is necessary to use specific selection pressures to verify their transgenic nature.

REFERENCES

1. BlackHall, Davey. M. R. and Power. J. B (1993) Isolation, culture and regeneration of protoplasts. In Plant Cell Culture: A practical approach. 2nd edition. Ed. Dixon. R. A. and Gonzales. R. A. Oxford University Press Inc. New York.
2. BlackHall, Davey. M. R. and Power. J. B (1993) Applications of protoplast technology. In : Plant Cell Culture : A practical approach. 2nd edition. Ed. Dixon. R. A. and Gonzales. R. A. Oxford University Press Inc. New York.
3. Dixon. R. A. (1985) Plant Cell Culture. A practical approach. IRL press, Oxford. Washington D. C.
4. Dixon. R. A. (1993) Transient gene expression and stable transformation. In : Plant Cell Culture : A practical approach. 2nd edition. Ed. Dixon. R. A. and Gonzales. R. A. Oxford University Press Inc. New York.
5. Franklin. C. I. and Dixon. R. A. (1993) Initiation and maintenance of callus and cell suspension cultures. In : Plant Cell Culture : A practical approach. 2nd edition. Ed. Dixon. R. A. and Gonzales. R. A. Oxford University Press Inc. New York.
6. Fromm. M., Taylor. L. P., and Walbot. V. (1985) Expression of genes transferred into monocot and dicot plant cells by electroporation. Proc. Natl. Acad. Sci. USA. 82 : 5824 - 5828.
7. Horsch. R.B, Fry, J., Hoffman, N., Neidermeyer, J., Rogers, S.O. Fraley, R.T (1989) Leaf disc transformation. In Gelvin, S. B. and Schilperoort (Eds), Plant Molecular Biology Manual. A5: 1 - 9.Kluwer Academic Publications. Derdrecht. Printed in Belgium.
8. Jefferson. R. A. (1987) Assaying chimaeric genes in plants.: the GUS gene fusion system. Plant Mol. Biol. Rep.5 : 387 - 405.
9. Klein, T.M., Wolf . E. D., Wu. R. and Sanford. J. C. (1987) High velocity microprojectiles for delivering nucleic acids into living cells. Nature. 327 :70 – 73.
10. Murashige. T. and Skoog. F. (1962) A revised medium for rapid growth and bioassays with tobacco tissue cultures. Physiologia Plantarum. 15 : 473 - 497.

Appendix - 1

GUIDELINES FOR THE PREPARATION OF A SCIENTIFIC RESEARCH ARTICLE

The data generated from the experiments can be compiled into a research article to be able to publish it in scientific journals. The article should ideally be written in a simple, direct style in third person and past tense. This compilation beginning with the title, author(s) and affiliation should conform to a general format consisting of the following: Abstract or summary, introduction, material and methods, results, discussion and literature cited (references).

Title

The title should be brief and convey precisely the nature of the experiment or the technique etc. If scientific names are used, it is essential to be very accurate about the nomenclature.

Author (s)

The name (s) of the author (s) should be given clearly in the specific format of the journal (some journals prefer that initials be provided instead of full names).

Affiliation

The complete address(es) of the author(s) should be given below the names. If the affiliations of the author differ, superscripts can be used to connect the correct affiliation with each author.

Abstract

The abstract is a concise summary of the experiment. It should contain the full information of the experiment including the conclusions drawn. The abstracts of articles accepted for publication are also printed in reference books and publications for easy access by other scientific workers. That is one reason that the abstract should be written in the third person. The information in the abstracts helps people to decide whether they need to request the author for a copy of the whole article.

Introduction

The information should explain clearly the reason for the investigation in the context of a few specific references of related work.

Material and methods

This should contain information regarding the experimental material (including the names of the species, cultivars, genotypes etc). The methods followed in the course of the experiment should be clearly explained in a simple manner. If procedures detailed in referenced papers were followed exactly, then it is adequate to cite those articles without further elaboration. Any variations to the generally accepted protocols should be indicated. The number of samples and replications used should be clearly stated.

Results

With few exceptions, all the data generated from the experiment should be represented in a tabular and / or figure form (graphs, diagrams) and all physical evidence of the results should be represented in the form of photographs. The titles for the tables and figures should be carefully chosen. The interpretation of the data and photographs if any should be very crisply presented without ambiguities.

The position for the tables and figures should be indicated in the margins of the results section, although the legends, tables and figures themselves are appended to the text, after the references.

Discussion

The results should be interpreted in the light of available literature and it is essential to quote and provide complete references wherever necessary.

Though modest conclusions can be made regarding the results obtained, it is always ideal to recommend further intensive study. Wherever references are cited, it is essential to follow the specific format of the particular journal. In some cases, the results and discussion can be combined and written together.

References

Only those references cited in the text should be given. Several accepted forms of literature citations are available and these vary with specific scientific journals. Therefore, the specific format should be applied.

Appendix - 2

ELECTRONICALLY ACCESSIBLE DATA FOR BIOTECHNOLOGISTS AND MOLECULAR BIOLOGISTS
(Sequence Database Searching and Some Interesting World Wide Web Sites)

Sequence database searching

Database searching can be carried out to search for sequences similar to the sequence that interests you. The databases can also be utilized to find out the sequence of specific genes which are utilized for vector construction. Any database sequence and its associated documentation can be retrieved and written into a personal file or a personal data library. The National Center for Biotechnology Information (NCBI) in the NIH (National Institute of Health web (www) and others can be accessed through the Genetics Computer Group (GCG). Programs like the BLAST (Basic Local Alignment Search Tool), Entrez and NLM PubMed (MEDLINE) can be used to search databases maintained at the NCBI over the internet.), Bethesda, Maryland, USA maintain several databases. There are several programs that can be used to search these databases and some of these can be searched through the world wide ograms like FastA, TFastA and several others can be used to access the GCG database.

Available databases

Several databases are maintained throughout the world: Gen Bank, PIR-Protein, SWISS-PROT, EMBL, GenEMBL (specifying both the GenBank and EMBL) etc. GenBank is the NIH genetic sequence database, which is an annotated collection of all publicly available DNA sequences.GenBank is part of the International Nucleotide Sequence Database Collaboration, which is comprised of the DNA DataBank of Japan (DDBJ), the European Molecular Biology Laboratory (EMBL), and GenBank at NCBI.These three organizations exchange data on a daily basis.

Most journals require submission of sequence information to a database prior to publication so that an accession number may appear in the paper. The methods of submission of sequences vary with different databases. The databases offer services like PubMed MEDLINE, the Entrez search system, BLAST sequence similarity searching and other E-mail servers. In addition to the DNA sequence databases, there are protein sequence databases like the SWISS-PROT and networks which can be used for the prediction of secondary structure of proteins like the Protein Design Group, EMBL, Heidelberg, Europe. HEADLINE can be used to search for research literature on any topic.

General NCBI information can be obtained over the internet (www) at info@ncbi.nlm.nih.gov and the NCBI home page comes up. The Entrez can be clicked on if the need is to find the sequence of a particular gene. Entrez needs information about the name of the gene etc to start the search. The sequence information can then be downloaded as needed. On the other hand if a search needs to be made to find out any similar sequences to a particular protein, the BLAST program is utilized. In this program, the choice can be made to locate protein-protein similarity by specifying blastp and swissprot or Gen EMBL. The similarity sequences in nucleotide databases can also be specified by choosing the options. The information can be downloaded and utilized. The databases of fasta or tfasta can be searched for similarity sequences through the GCG programs. The GCG program is available only on the Wisconsin package. This runs on UNIX and one can logon to the system over the network by using a terminal emulation software eg.: EM340.

Interesting WWW sites

There are several interesting **World Wide Web** sites that can be explored by a scientific worker in the areas of Molecular Biology and Biotechnology. Some of these are given below.

Type out **http://** before every address.

- webcrawler.cs.washington.edu/webcrawler/webquery.html
  [*For general information.*]

- www.public.iastate.edu/~pedro/research_tools.html
  [*Molecular biology info.*]
- macserver.molbio.gla.ac.ukbest.gdb.org/best.html
  [*List of researchers in Molecular Biology.*]
- www.atcg.com/atcg
  [*Info on restriction enzymes.*]
- ***alpha.genebee.msu.su:80/supplier***

  or

  web.frontier.net/MEDMarket/indexes/indexmfr.html
  [*Biotech companies and suppliers of reagents and chemicals.*]
- www.atcc.org/atcc.html
  [*American type culture collection of plasmids and strains.*]
- cgsc.biology.yale.edu/top.html
  [*Database of E. coli strains & genes.*]
- Yahoo. Com. Science search.

Appendix - 3

CELL AND TISSUE CULTURE NEEDS

Table 1: Composition of some commonly used tissue culture media.

	MS	B_5	Nitsch's	N_6
	[Amount (mg / L)]			
Macro nutrients				
$MgSO_4.7H_2O$	370	250	185	185
KH_2PO_4	170	--	68	400
$NaH_2PO_4.H_2O$	--	150	--	--
KNO_3	1900	2500	950	2830
NH_4NO_3	1650	--	720	--
$CaCl_2.2H_2O$	440	150	--	166
$(NH_4)_2.SO_4$	--	134	--	463
Micro nutrients				
H_3BO_3	6.2	3	--	1.6
$MnSO_4.4H_2O$	22.3	--	25	4.4
$MnSO_4.H_2O$	--	10	--	3.3
$ZnSO_4.7H_2O$	8.6	2	10	1.5
$Na_2MoO_4.2H_2O$	0.25	0.25	0.25	--
$CuSO_4.5H_2O$	0.025	0.025	0.025	--
$CoCl_2.6H_2O$	0.025	0.025	0.025	--
KI	0.83	0.75	--	0.8
$FeSO_4.7H_2O$	27.8	--	27.8	27.8

	MS	B_5	Nitsch's	N_6
	[Amount (mg / L)]			
Na_2EDTA. $2H_2O$	37.3	--	37.3	37.3
EDTA Na Ferric	--	43	--	--
Sucrose (g)	30g	20	20g	50g
Organic supplements (Vitamins)				
Thiamine HCl	0.5	10	0.5	1
Pyridoxine HCl	0.5	1	0.5	0.5
Nicotinic acid	0.5	1	5	0.5
Myoinositol	100	100	100	--
Glycine	2	--	2	--
Folic acid	--	--	0.5	--
Biotin	--	--	0.05	--
PH	5.8	5.5	5.8	5.8

- Phyta-agar or phyta-gel is added for solidification of media.
- Growth regulators are not included since they can differ for different plants.
- MS medium (Murashige and Skoog 1962); B5 medium (Gamborg et al 1968).
- Nitsch's medium (Nitsch and Nitsch 1969); N6 medium (Chu 1978).

Table 2. Some surface sterilizing solutions used for plant tissue culture:

Disinfectant	Concentration (%)	Exposure time (min)	Remarks
Calcium hypochlorite	9-10 5-30		
Sodium hypochlorite	0.5-1.5	5-30	
Hydrogen peroxide	3-12	5-15	
Ethyl alcohol	70-95	0.1-5.0	
Bromine water	1-2	2-10	
Silver nitrate	1.0	5-30	
Mercuric chloride	0.1-1.0	2-10	Most frequently used
Antibiotics	4-50 mg l^{-1}	30-60	
Commercial bleach	10-20	5-30	Contains 5% sodium hypochlorite

Table 3: Protoplast culture medium:

Mineral salts	Amount mg/L
$NH_4\ NO_3$.	600.00
KNO_3.	1900.00
$Ca\ Cl_2 . 2H_2O$	600.00
$Mg\ SO_4 . 7H_2O$	300.00
KH2 PO_4.	170.00
KCl	300.00
Sequenstrene 300 Fe	28.00
KI	0.75
$H_3\ BO_3$.	3.00
$Mn\ SO_4 . H_2O$	10.00
$Zn\ SO_4 . 7H_2O$	2.00
$Na_2\ Mo\ O_4 . 2H_2O$	0.25
$Cu\ SO_4 . 5H_2O$	0.025
$Co\ Cl_2 . 6H_2O$	0.025
Sugars:	
Glucose	68400.00
Sucrose	125.00
Fructose	125.00
Ribose	125.00
Xylose	125.00
Mannose	125.00
Rhamnose	125.00
Cellobiose	125.00
Sorbitol	125.00
Mannitol	125.00
Organic acids: (of pH 5.5 with $NH_4\ O_4$)	
Sodium pyruvate	5.0
Citric acid	10.0
Malic acid	10.0
Fumaric acid	10.0
Vitamins:	
Inositol	100.0
Nicotinamide	1.0

Mineral salts	Amount mg/L
Pyridoxine Hcl	1.0
Thiamine Hcl	10.0
D-Calcium pantothenate	0.5
Folic acid	0.2
P-Aminobenzoic acid	0.01
Biotin	0.005
Choline chloride	0.5
Riboflavin	0.1
Ascorbic acid	1.0
Vitamin A	0.005
Vitamin D3	0.005
Vitamin B12	0.01
Optional additions:	
Growth regulators (2,4 - D, Zeatin, NAA, Kinetin, BAP etc).	
Casamino acid	
Coconut water.	

Appendix - 4

PREPARATION OF PHENOL*

SATURATION OF PHENOL:

- **Add the following to a 500 g bottle of phenol**

  130 ml distilled water

  6ml 1 M tris pH 7-8

  5 ml 3 M NaOH

  0.5 g hydroxyquinoline

- **T E buffer**

  10 mM Tris-HCl

  1 mM EDTA.

  Adjust the pH to 8.0.

Let stand overnight in hood with top loosened. Add 50 ml TE (10:1)buffer so that an upper aqueous layer appears. Store in a brown bottle in the fridge.

(Caution ! Use gloves !)

WATER-SATURATION OF PHENOL

Thaw crystals of phenol at 65^0C in a water bath and mix one part of phenol and one part of sterile distilled water. Mix well and allow two phases to separate. Store at 40^0C.

BUFFER-SATURATION OF PHENOL

Prepare water saturated phenol (Thaw phenol crystals at 65° C in a water bath and mix one part of phenol and one part of sterile distilled water. Mix well and allow two phases to separate.). Take an aliquot of the water - saturated phenol and add an equal volume of 0.5 M Tris-HCl (pH 8.0). Mix well and let the phases separate at 15 to 30^0 C. Transfer the upper aqueous phase to another tube. Add equal volumes of 0.1 M Tris-HCL (pH 8.0) and repeat the above step. Check the pH of the aqueous phase and repeat until the pH is 8.0.

Appendix - 5

SUGGESTED READING

1. Ammirato, P.V (1984): In. I. K. Vasil (Ed): Cell Culture and Somatic Cell Genetics of plants. Vol I: 139-151 (*Tissue Culture*).
2. Anderson, W.F and Diacuma Kos E.G (1981): Sci Am, 245 (i): 106 - 121 (*Genetic Engineering*).
3. Andrews, A.T (1986): Electrophoresis Theory, techniques and Biochemical and Clinical applications. Oxford Univ. Press. Oxford.
4. Aneja K.R (1993): Experiments in *microbiology*, *plant pathology* and *tissue culture*. Wishwa Prakashan, New Delhi.
5. Bajaj, Y.P.S(ed) (1990): *Biotechnology* in agriculture and forestry vol. 1-11 Springer - Verlag. Berlin.
6. Bapat et al (1987): Plant cell Rep 6: 393 - 395 (*Tissue Culture*).
7. Beckman, J.S and M. Soller (1986): In *Plant molecular* and *Cell Biology*. Oxford surveys 3:196 - 205.
8. Bevan M. (1984) Nucleic acids Res 12: 8711 - 8721 (*plant transformation*).
9. Bevan M et al (1983) Nature (Lond) 304: 184 - 187 (*Plant transformation*).
10. Bhojwani, S.S (1990): *Plant tissue culture*: Applications and limitations Elsevier, Amsterdam.

11. Biotol book on Biotechnological innovations in crop improvement. Open University Publication Netherlands and Thames Polytechnic U.K (*Biotechnology*).

12. Brown, T.A (1990): Gene cloning. Chapman & Hall, London (*Molecular Biology*).

13. Carlson, P.S et al (1972): Proc Nat. Acad Sci USA 69:2292 - 2294 (*parasexual inter specific hybridisation*).

14. Chapman, J.M and Avery G (1981): The use of *Radio isotopes* in life sciences. George Allen and Unwin London.

15. Chowdhury, M.K.U and I.K. Vasil (1992): Plant cell Report 11:494 - 498. (*Direct gene transfer - micro projectiles*).

16. Cocking, E.C (1960): Nature (Lond) 187:927 - 929 (*Isolation of protoplasts*).

17. Cocking, E.C and Davey M.R (1987): Gene transfer in cereals. Science 236:1259 - 1262.

18. Cohen et al (1972) PNAS USA: 69:2110 (*Culture of bacteria - transformation*).

19. Colowick, S.P and Kaplan N.O (Eds): Methods in *enzymology*. Acad press, New York.

20. Cooper, T.G (1977): Tools of *Biochemistry*. John Wiley & Sons, New York. Ch. 1.

21. De Block, M et al (1987) EMBO. J: 6(9):2513 - 2518 (*Engineering herbicide resistance in plants*).

22. De. K.K (1992): An introduction to *plant tissue culture*. New Central book agency, Calcutta.

23. Dellaporta *et al* (1983): Plant Molecular Biology Rep. 1(4):19 (*Isolation of DNA*).

24. Del Sal, G et al (1989): Biotechniques 7:514 - 519 (*Isolation of DNA - CTAB method*).

25. Ford, T.C and Graham J.M (1991): An introduction to *centrifugation*. Bios Scientific Publishers Ltd. Oxford.187

26. Fraley R.T et al (1983): Proc. Natl Acid Sci. USA 80:4803 - 4807 (*Expression of bacterial genes in plant cells*).

27. Frearson E.M et al (1973): Dev. Biol. 33:130 - 137 (*Protoplast culture*).

28. Fredric et al (1993): Current protocols in *molecular biology* 1:7.0.3 - 7.7.22.

29. Fromm, M.E., Taylor, L.P., Walbot, V. (1986) Nature (Lond) 319:791 - 793 (*Gene transfer - electroporation*).

30. Gamborg O.L et al (1968): Exp cell Res. 50:151 - 158 (*Tissue culture*).

31. Gamborg O.L et al (1983): Plant cell Rep. 2:209 - 212 (*Tissue culture*).

32. Ganapathi et al (1992): Plant cell Reports 11:571 - 575 (*Tissue culture*).

33. Giri C.C and G.M.Reddy (1994): Current science 67(7):542 - 545 (*Synthetic seeds*).

34. Griffith, O.M (1983): Techniques in preparative, zonal and continuous flow *ultra centrifugation*. 4th ed. Beckman Instruments Inc. Palo Alto C.A.

35. Guha S and Maheshwari S.C (1964): In vitro production of embryos from anthers of *Datura*. Nature (Lond) 204:497.

36. Hames, B.D and Rick wood. D (1990) (Ed): Gel electrophoresis of proteins - A practical approach. 2nd Edn. IRL press Oxford (*electrophoresis - useful for research workers*).

37. Hames B.D and S.J Higgins (1985): (Ed) Nucleic acid hybrdization, a practical approach. IRL press.

38. Hancock W.S (Ed) (1990): *HPLC* in biotechnology. Wiley Inter Science, New York.

39. Herrera - Estrella et al (1983): Nature (Lond) 303:209 - 213 (*Gene transfer through Ti-plasmid vector*).

40. Hester R.E and Girling, R.B (Eds) (1991): *Spectroscopy* of biological molecules. Royal Society of Chemistry. Special publication, Cambridge.

41. Horsch. R.B et al (1985): Science 227:1229 - 1231 (*Gene transfer - Agrobacterium system*).

42. Horsch. R.B, Fry, J., Hoffman, N., Neidermeyer, J., Rogers, S.O. Fraley, R.T (1989) Leaf disc transformation, in Gelvin, S. B. and Schilperoort (Eds),*Plant Molecular Biology* Manual. A5: 1 - 9. Kluwer Academic Publications.Derdrecht-Printed in Belgium.

43. Innis, M.A et al (1988): Proc. of Natl. Acad. of Science 85(24):9436 - 9440.

44. Jayaraman J (1981): Laboratory, manual in *Biochemistry*. Wiley Eastern Ltd.

45. Jefferson, R.A (1987): Plant Molecular Biology Rep. 5:387 - 405 (*GUS assay*).

46. Jena, K.K et al (1990): Theor. App. Genet. 80:737 - 745 (*RFLPs*).

47. Jena, K.K et al (1992): Theor, Applied Genet 84:608 - 616 (*RFLP mapping*).

48. Jena, K.K et al (1994): Genome 37:382 - 389 (*RFLP mapping*).

49. Jones and Richards (1990): *Practical Genetics*.

50. Joshi P. (1999): Genetic Engineering and Its Applications, Agro Botanica, Bikaner.

51. Kao, K.N and Michayluk. M.R (1989): In Y.P.S Bajaj (ed) Biotechnology in Agriculture and Forestry. Vol. 8. 277 - 288 (*protoplasts - Genetic Engineering*).

52. Khorana, H.G (1979): Science 203:614 - 625 (*Synthesis of a gene*).

53. Klein, T.M., Wolf . E. D., Wu. R. and Sanford. J. C. (1987) High velocity microprojectiles for delivering nucleic acids into living cells. Nature 327 :70 - 73 (*Micro projectiles - gene transfer*).

54. Klein, T.M et al (1988 a): Proc. Natl. Acad. Sci. USA 85:4305 - 4309 (*Gene transfer using micro projectiles*).

55. Klein, T.M et al (1988 b): Biotechnology 6:559 - 563 (*Gene transfer using micro projectiles*).

56. Kumar U. (1999): Methods in Plant Tissue Culture. Agro Botanica, Bikaner.

57. Kumar U. (1999): Synthetic Seeds for Commercial Crop Production, Agro Botanica, Bikaner

58. Lacey, A.J (Ed) (1989): *Light microscopy* in Biology - A practical Approach. IRL press. Oxford.

59. Larkin, P.J and Scowcroft W.R (1981): Theor Appl. Genet. 60:197 - 214 (*Somaclonal variation*).

60. Lindsey, K et al (1991): In Plant Tissue Culture Manual. Kluwer Academic Publishers, Netherlands. (*Transformation by Agrobacterium system*).

61. Mahadevan A and Ulaganathan (1992): Techniques in *molecular plant pathology*. Sivakami Publications, Madras.

62. Mathur et al (1989): Plant Science 60:111 - 116 (*Synthetic Seeds*).

63. Mc Cormick. S (1991): In plant Tissue Culture Manual B6. 1 - 9. (*Agrobacterium mediated transformation*).

64. Mc Innes. E et al (1991): Plant cell Report 9:647 - 650 (*Agrobacterium rhizogenes - hairy root production*).

65. Mc Clintock. B (1957): Cold Spring Symposium on Quantitative Biol. 21:197 - 216 (*Transposable elements*).

66. Micklos, D.A and Freyer, G.A (1989): Biotechnology Education vol. I, No. 1, P 16 - 22. Pergamon Press Great Britain.

67. Mukhopadhyay. A et al (1992): Plant cell Reports 11:506 - 513 (*Agrobacterium mediated gene transfer*).Mullis. K.B (1990): Scientific American (*The polymerase chain reaction*).

68. Mullis. K.B and Faloona. F.A (1987): Meth Enzymol 155:335 - 350.

69. Murashige T and Skoog F (1962): Physiologion plantarum 15:473 - 497.

70. Murray and Thomson (1980): Nucl. Acid Res 8:4321.

71. Murray and Thomson (1990): Molecular plant development.

72. O.W, et al (1986): Science 234:856 - 859 (*Luciferase gene*).

73. Parrot W.A et al (1989): Plant cell, Tissue and Organ Culture 16:15 - 21. (*Somatic embryogenesis - Tissue Culture*).

74. Patterson A.H et al (1988): Nature 335:721 - 726 (*RFLP mapping*).

75. Plummer, D.T (1977): An introduction to practical *biochemistry*. Tata - Mc Graw Hill, Bombay.

76. Potrykus. I et al (1985): Mol. Gen. Genet. 199:183 - 188 (*Direct gene transfer into monocot plant cells*). 189Power, J.B and Cocking, E.C (1968): Biochem. J. 111:33 (*protoplast isolation*).

77. Power, J.B et al (1970): Nature (Lond) 225:1016 - 1018 (*protoplast fusion*).

78. Purohit, S. S. (1999): Biotechnology: Fundamentals and Applications, III ed. Agro Bios (India), Jodhpur.

79. Razdan M.K (1993): An introduction to *plant tissue culture*. Oxford IBH Publishing Co. Pvt. Ltd.

80. Reddy G.M (1988): In Nature of genetic variation in maize. Ed. Chauhan P.S and P.P. Reddy EMSI. BARD, Bombay pp. 193 - 200.(*Application of molecular biology techniques*).

81. Reddy G.M (1995): Laboratory manual on Basic Techniques in Plant Molecular Biology. Published by the Centre for Plant Molecular Biology. Dept. of Genetics, Osmania University, Hyderabad, India.

82. Redenbaugh. K (1990): Hort Sci. 25:225 - 251 (*Synthetic seeds*).

83. Redenbaugh. K et al (1986): Biotechnology 4:797 - 801 (*Synthetic seeds - encapsulation of somatic embryos*).

84. RFLP Training Course Laboratory Manual (1989): Rocke Feller Foundation. Cornell University, USA.

85. Rhodes et al (1988): Science 240:204 - 207 (*Genetic transformation from protoplasts*).

86. Saghai - Maroof et al (1984): Proc. Natl. Acad. Sci. 81:8014 - 8018 (*Method of isolation of DNA*).

87. Saiki, R. K et al (1985): Science 230:1350 - 1354 (*PCR*).

88. Saiki, R.K et al (1988): Science 239:487 - 489 (*PCR*).

89. Sambrook J. et al (1989): *Molecular cloning* - A Laboratory manual 2nd Ed. Cold Spring Harbor Laboratory press Cold Spring Harbor, New York.

90. Schuler M.A and Zielinski R.E (1990): Methods in plant *molecular biology*. Acad Press Inc.

91. Scott, O.R and A.J Bendich (1988): *Plant Molecular Biology* Manual. A6:1 - 10. Kluwer Academic Publishers, Dordrecht. Printedin Belgium.

92. Segel I.H (1976): *Biochemical Calculations*. 2nd Ed. John Wiley and Sons, New York.

93. Sharpe P.T (1988): Laboratory techniques in Biochemical and Molecular biology Vol. 8. *Methods of separation*. Eds. Burden R.H and van Knipenberg. P.H. Elsevier Amsterdam. New York. Oxford.

94. Shaw, C.H (1990) Ed: *Plant Molecular Biology*.

95. Shaw, C.H et al (1983): Gene 23:315 - 330 (*Gene transfer to plants*).

96. Skoog D and White D (1982): Fundamentals of Analytical chemistry. 4th Ed. Saunders College Publishing Co.PA, USA.

97. Slater R.J (Ed) (1986): Experiments in Molecular Biology. Humana Press. Totawa, N.J, USA.(*Protocols for under graduates*).

98. Sokal R.R and Rotilf F.J (1987): Introduction to *Biostatistics*.2nd Ed. W.H Freemand and Co.New York,USA.

99. Somers D.A (1992): Biotechnology vol. 10. 1589 - 1594 (*Transgenic Oat plants*).

100. Southern E.M (1973): J. Mol Biol. 98:403 - 517 (*Southern blotting*).

101. Stansfield W.D (1991): Theory and problems of *Genetics*. 3rd Ed. Schaun's Outline Series. Mc -Graw Hill Book Co.

102. Stryer L (1988): *Biochemistry*. 3rd Ed. W.H Freeman and Co., New York, USA.

103. Tanksley S.D et al (1989): *Biotechnology*. 7:257 - 264 (*RFLP mapping*).

104. Twell D et al (1991): *Plant Tissue Culture manual.* D2. 1 - 14. Kluwer Academic Publishers. Printed in Netherlands.

105. Visser R.G.F (1991): *Plant Tissue Culture* manual B5. 1 - 9. Kluwer Academic Publishers. Printed in Netherlands.

106. Walker J.M (Ed) (1984): Methods in *Molecular Biology* Vol. 1 - 4. Humana Clifton N.J, USA.

107. Watson J.D et al (1983): *Recombinant DNA* - A short course. W.H Freeman, New York.

108. Wedgewood M (1989): Tackling Biology *Projects*. Macmillan Education Ltd, Basing Stoke.

109. Wharton D and Mc Carty R (1972): Experiments and methods in *Biochemistry*, Macmillan, New York. 5 - 20.

110. White F.F et al (1985): J. Bact. Biol. 164:33 - 44 (*Agrobacterium - rhizogenes gene transfer*).

111. Williams J.G and R.K patient. *Genetic Engineering*. IRL Press Oxford. Washington - DC, USA.

112. Wilson K and J Walker (Ed) (1994) Practical *Biochemistry*. 4th Ed. Cambridge University Press, U.K.